Barriers to International Technology Transfer

NATO ASI Series

Advanced Science Institutes Series

A Series presenting the results of activities sponsored by the NATO Science Committee, which aims at the dissemination of advanced scientific and technological knowledge, with a view to strengthening links between scientific communities.

The Series is published by an international board of publishers in conjunction with the NATO Scientific Affairs Division

A Life Sciences **B Physics**	Plenum Publishing Corporation London and New York
C Mathematical and Physical Sciences **D Behavioural and Social Sciences** **E Applied Sciences**	Kluwer Academic Publishers Dordrecht, Boston and London
F Computer and Systems Sciences **G Ecological Sciences** **H Cell Biology** **I Global Environmental Change**	Springer-Verlag Berlin, Heidelberg, New York, London, Paris and Tokyo

PARTNERSHIP SUB-SERIES

1. **Disarmament Technologies**	Kluwer Academic Publishers
2. **Environment**	Springer-Verlag / Kluwer Academic Publishers
3. **High Technology**	Kluwer Academic Publishers
4. **Science and Technology Policy**	Kluwer Academic Publishers
5. **Computer Networking**	Kluwer Academic Publishers

The Partnership Sub-Series incorporates activities undertaken in collaboration with NATO's Cooperation Partners, the countries of the CIS and Central and Eastern Europe, in Priority Areas of concern to those countries.

NATO-PCO-DATA BASE

The electronic index to the NATO ASI Series provides full bibliographical references (with keywords and/or abstracts) to more than 50000 contributions from international scientists published in all sections of the NATO ASI Series.
Access to the NATO-PCO-DATA BASE is possible in two ways:

– via online FILE 128 (NATO-PCO-DATA BASE) hosted by ESRIN, Via Galileo Galilei, I-00044 Frascati, Italy.

– via CD-ROM "NATO-PCO-DATA BASE" with user-friendly retrieval software in English, French and German (© WTV GmbH and DATAWARE Technologies Inc. 1989).

The CD-ROM can be ordered through any member of the Board of Publishers or through NATO-PCO, Overijse, Belgium.

Series 4: Science and Technology Policy – Vol. 11

Barriers to International Technology Transfer

edited by

John Kirkland
National Institute of Economic and Social Research,
London, United Kingdom

Kluwer Academic Publishers

Dordrecht / Boston / London

Published in cooperation with NATO Scientific Affairs Division

Proceedings of the NATO Advanced Research Workshop on
Barriers to International Technology Transfer
London, United Kingdom
September 17–20, 1995

A C.I.P. Catalogue record for this book is available from the Library of Congress

ISBN 978-90-481-4781-6

Published by Kluwer Academic Publishers,
P.O. Box 17, 3300 AA Dordrecht, The Netherlands.

Kluwer Academic Publishers incorporates the publishing programmes of
D. Reidel, Martinus Nijhoff, Dr W. Junk and MTP Press.

Sold and distributed in the U.S.A. and Canada
by Kluwer Academic Publishers,
101 Philip Drive, Norwell, MA 02061, U.S.A.

In all other countries, sold and distributed
by Kluwer Academic Publishers Group,
P.O. Box 322, 3300 AH Dordrecht, The Netherlands.

Printed on acid-free paper

TABLE OF CONTENTS

TECHNOLOGY TRANSFER IN THE TRANSITION ECONOMIES: CASE STUDIES

CONCLUSIONS

Preface

The importance of technology transfer to innovation and wealth creation is now recognised by most governments. As the policy debate has intensified, however, it has become clear that the problem of encouraging successful transfer is complex, and requires an interdisciplinary approach.

The collection of papers in this volume is deliberately diverse. It offers perspectives from economics, sociology, science, engineering and public administration, and also from outside academic life, from those involved at the 'sharp end' of technology licensing and administering government research programmes. Contributions are also drawn from a range of national backgrounds—the authors are drawn from ten countries, from throughout Europe and North America.

The main focus for the papers was a NATO Advanced Study Workshop, which took place at the National Institute of Economic and Social Research, London, in September 1995. Unfortunately time and space has prevented all of the contributions appearing here, but all those who attended played an important role in making the event such a success. Thanks are also due to Dr Alain Jubier and his colleagues at NATO, without whose support and advice the seminar could not have take place, to my fellow organising committee members Dr Katalin Balázs, Dr Linda Parker and Professor Steve Woolgar, and to Monica Miglior who, in addition to assisting in the conference organisation, provided detailed notes on sessions which helped greatly in later analysis.

Another vital factor in the success of the seminar was the help and assistance of my colleagues at the National Institute. Particular thanks are due to Fran Robinson, whose patience, determination and attention to detail has ensured the production of this volume, to Barbara Daly and Michele Ockenden for excellent secretarial support, and to Gill Clisham, Ann Hall, Pat Shaw and Jean MacRae, for services beyond the call of duty which ensured that our guests were provided with the friendly welcome for which the Institute is renowned. I hope that the group will have further opportunities to meet in the future and that, in the meantime, the present volume will prove thought provoking for academics, practitioners and policymakers.

JK

July 1996

INTRODUCTION: THE PROBLEM OF TECHNOLOGY TRANSFER

JOHN KIRKLAND
National Institute of Economic and Social Research
2 Dean Trench Street, London SW1P 3HE

1. What is Technology Transfer?

Technology transfer is the process by which ideas and techniques generated in one place find an application in another. In the chapters of this book, emphasis is placed on transfer between the public sector science base, most typically universities, and industry. This can be justified by the degree of government investment in science and technology policy, and expectation that it will lead to public benefit. Yet it should be recognised that technology transfer from universities represents only one aspect of a much broader process. The flow of ideas between and within private sector organisations, although less visible and accessible to the researcher, will be of much greater volume. This is largely beyond the scope of our book.

Within the field of academic–industry relations, the mechanisms for technology transfer are more numerous than is often imagined. They embrace formal licensing agreements, collaborative research, consultancy arrangements, equipment sharing, short courses and, increasingly, specialised spin-off companies and science park initiatives. Most have grown significantly during the past decade. Moreover the traditional teaching function of universities, insofar as it provides large numbers of graduates who will themselves take new ideas into industry, could also be seen as part of the process. This wide range of mechanisms has led some commentators to develop a wider concept of 'knowledge' rather than 'technology' transfer [11].

As Todorov points out in his chapter, knowledge transfer itself has several definitions. His own selection comes from Rogers, who refers to 'a process thanks to which an innovation reaches members of a social system in time through communication channels'. Once again, the talk is of a process rather than a single act by which knowledge passes from instigator to user. There has been increasing recognition in recent years that transfer can no longer be seen as a 'linear process', which Schumpeter defined as comprising invention, innovation and diffusion, but is better viewed as a two-way channel of communication. This might not involve any formal exchange of intellectual property, but the passing of concepts, know-how and ideas across a much wider range of actors. For example, it may involve ideas and expertise moving from industry to academics, as well as from academic 'inventors' to industrial and other 'users'.

The concept of a continous and two-way process also recognises that the innovation process is often based on incremental (or, as Todorov states, cumulative) change, rather than major new inventions. Attempts to categorise types of innovation go back many years. More recently, Betz has spoken of four categories, each based on level of impact—radical comparative innovations, incremental innovations to existing technologies, radical systems innovations and next generation technology innovations [2]. The distinction is a valuable one for policy purposes, but it is clear that the majority of new ideas come into the second category. Likewise, most of the information or material being exchanged is neither new or capable of additional legal protection; it may simply involve the spreading of existing techniques, or their application to a new problem or business environment.

That the process has received such attention in recent years is testament to its perceived importance to economic prosperity. Before detailing the findings of our conference participants in detail, however, this introductory section introduces the main actors in the transfer process, and provides an overview of the most commonly cited barriers to its efficient operation.

2. Participants in the Technology Transfer Process

For the transfer of ideas to take place in any systematic way, participants must have the incentives to meet and exchange knowledge, and the confidence that their participation will be beneficial. Essential to this confidence is the belief that, in the unusual event that cooperation leads to a major breakthrough, recognition and reward will be available to all parties. There is some similarity to a syndicate buying tickets for a national lottery (an experience to which the UK is just becoming accustomed). The prospects of winning the jackpot are remote, but partners consider it worthwhile to put in place mechanisms to cater for the eventuality should it arise, often by way of legally enforceable agreements to distribute the winnings in a particular manner. Such agreements may not, in themselves, provide a significant incentive to take part, but their absence (or the lack of confidence in them) can cause considerable difficulties.

The process of technology transfer thus relies heavily on the individual. In the words of a former senior adviser to the UK Department of Trade and Industry and senior industrialist, speaking to a recent seminar:

> 'When I saw the title to your Seminar Series, 'The Barriers to Technology Transfer, I thought this could be a very short series because the only barriers to technology are people and everything else is fairly easy. Whether it's people as individuals, people as collective organisations, or organisations as a set of interrelated bodies, a lot of the problems of technology transfer do come down to people and social things. In fact, one of the reasons I became associated with the [Economic and Social Research Council] was that in my experience in industry over the past twenty years—even in an industry which is an archetypical high-tech industry—it is still true that most of the problems that we wrestle with as industrial managers are social-related problems and not technological-related problems.'

Thus a 'barrier' to technology transfer can be seen as a factor which prevents individuals (or the organisations for which they work) participating fully in the process described above, or, worse still, leads them to obstruct that process. In some cases, such obstacles can be overcome through government intervention and financial initiatives, but even these will be of limited usefulness if the motives of individuals are not fully in sympathy with those of funders; if, for example, the individual academic places more emphasis on securing his next major research grant than on communicating or exploiting the results of the last one. In any event, given the breadth of the transfer process described above, it is inconceivable that direct government funding can support more than a small minority of the necessary interaction. More likely, it can act as a catalyst by sending suitable signals and creating incentives, and by the removing of unnecessary barriers.

The response of academic communities to overtly industrial or commercial research and development was one of many areas identified in our conference in which the experience of the 'transition' economies of central and eastern Europe, although on a different scale to those of Western participants, is by no means unrecognisable. Thus the description by a Hungarian delegate of 'perception gaps' and the feeling, ingrained in Hungarian academics over many years, that 'business is dirty', provoked many wry smiles in the audience.

From the study by Webster and Packer we see that, despite generous financial incentives, attitudes remain ambiguous. For many academics, the terms of their contracts with universities allow for at least half, and sometimes more, of the revenue gained from inventions made during the course of their employment to be retained personally, without the risk of patenting or defence costs. The personal benefits of a new research project, by contrast, are often confined to prestige, the possibility that promotion prospects (often within a relatively narrow salary band) will be enhanced, and the ability to keep a research team together. Yet this long-standing provision—quite unlike the terms offered to scientists in industry—does not seem to be a major motivating factor.

The apparent reluctance—at least by academics—to respond to direct financial incentives, leads us to consider the personal motives of those involved in the technology transfer process in some detail. Identification of these individuals is itself a subject of study; some of the recent work is referred to in the chapter by Rahm. At this stage, however, we can group them into four categories—university academics, university administrators, industrial scientists and industrial administrators.

Of these four, university academics have received the most attention in the current literature. Academics, of course, are themselves a diverse group. The new, entrepreneurial attitudes of the last ten years have not affected all university staff, many of whom took their first university post in a very different environment. Even those who can be described as entrepreneurial may be sub-divided. Etzkowitz [6] identifies the roles of mogul, sustainer, adviser and craftsperson academics, all of whom are pursuing different objectives.

We can, however, advance some general comments. Several papers to our conference confirmed the UK experience that peer group recognition is likely to weigh more heavily with the individual academic than any conflicting steer from university administrators or government. In recent years, also, academics have suffered from increasing demands on their time, both from increasing student numbers and administrative load. Thus there is a growing notion of opportunity cost to consider in allocating time. This is reinforced by the way in which performance is evaluated by government. In the United Kingdom, this relies largely on a four-yearly research assessment exercise which, despite some change in recent years, still stresses peer review, publication record and levels of research grant income more heavily than subsequent usefulness. Given this, and the risks involved in devoting scarce time and resources to transfer activities, the academic whose chief concern at the end of a major project is to publish the results and secure his next external grant, might be thought to be behaving entirely rationally.

In the past fifteen years, most UK universities have also developed administrative systems to assist in the process of research management and industrial liaison. It would be wrong to suggest that these have taken the process of knowledge transfer out of the hands of academics. In most cases the offices concerned are small—unpublished surveys by the former University Directors of Industrial Liaison (UDIL) organisation suggest 4–8 staff to be typical—and staffed largely by generalists rather than specialists in any specific area. Their precise role varies, but often incorporates both promotional activity—such as publicising research capabilities, mounting research applications and seeking licences for commercial exploitation—with a more defensive, legal role, including contract negotiation, costing of projects and ensuring that the university receives adequate returns on its activity. The potential of these offices to act as a catalyst for successful university–industry liaison has been recognised by central government. In 1992 the Department of Trade and Industry introduced two competitive funding schemes, one to facilitate extra staff in such offices, the other to allow them to employ external consultants to conduct technology audits of their institutions.

The motivation for industrial liaison staff can be as complex as that for academics. Most are employed as part of the central university administration, and it is therefore possible that their objectives will conform more closely to those of their institutions than those of academics, who might consider themselves more closely allied with a subject peer group. But institutional objectives towards technology transfer are themselves complex and, partly as a result, their means of evaluating the success of administrators vary. Some still regard the activity as unlikely to produce tangible benefit in the short term, or accept that the marginal contribution of administrators, as opposed to academics, cannot be isolated. This approach allows the Industrial Liaison Officer considerable freedom to develop informal links, but is more difficult to justify, in the short term at least, on strictly financial grounds. Others will consider the defensive or regulatory role of such offices to be paramount, and judge them a success if the university avoids losses or legal difficulties in the area of research.

These distinctions are important because the industrial liaison function, like other areas of university life, are increasingly subject to performance indicators. These in turn demand accurate comparisons between universities. It must be said that the means of comparing performance currently available are not likely to lead to an emphasis on exploitation. At present, the two most commonly used criteria are gross research income and recovery rates of indirect costs (or 'overheads'). The former case relies much more heavily on research councils and government, which still provide the majority of funding for UK academic research; the latter leads to the danger that university–industry negotiations will concentrate on the costs of collaboration, rather than on the potential benefits. It is certainly true that costing and pricing policy has occupied a disproportionate share of the debate on collaboration in recent years. By contrast, UK universities are not required to publish their level of income from commercial exploitation or licensing of intellectual property as a discrete item—it can simply be reported as 'Other Income', in which it is mixed with several other headings.

The role of scientists in industry also deserves further attention. Despite the growth of technology management as a discipline in recent years, and attempts to improve the prospects of those with a technological background reaching senior decision-making levels, many industrial scientists retain an affinity with their subject 'peer group' (including scientists working in similar fields in universities) as well as their employers. Many industrial scientists have—and some actively seek—an active publications record. Contact is also facilitated through professional bodies, which reinforce the importance of the 'chemist' and the 'engineer' regardless of employing organisation. As Robinson pointed out, informal relations of this nature can be vital in overcoming the legal and financial obstacles to technology transfer which entrenched institutional positions sometimes promote. There may be occasions when the scientists in university and company identify more common interest with each other than with their colleagues in finance, legal and other managerial roles. Industrial R&D personnel themselves may also be divided, for example between laboratories at corporate and divisional level. Betz [2] notes that 'when a diversified firm adopts such a mixed form (having both a corporate research lab and divisional labs), corporate management can still expect the two forms of research organisations to 'snipe' at each other.'

There is also evidence of industrial scientists having to acquire greater negotiating skills and managerial ability within their own organisations. In the UK, the role of central R&D departments appears to be changing, partly as a result of increased decentralisation and the recession of the early 1990s. In larger companies, such departments have increasingly had to secure 'internal customers' for their projects, rather than relying on core funding from the centre. Perel [14] and Corcoran [4] describe similar trends in the United States, arguing that research managers can no longer work in isolation, but need to develop the capacity to interact with other levels in their own organisations. The long-term effects of these changes on the importance attached to research, technology and intellectual property remain to be seen. Whilst the greater integration of such issues into company planning and budgeting might increase their profile, it is also possible that the rise of the customer–supplier relationship has left research and development more exposed to short-term crises.

Less is known about the individuals who administer technology transfer and intellectual property matters within companies than in universities. In the UK, only the largest companies, or those whose sector demands a particular interest in academic research, such as pharmaceutical companies, have established academic liaison offices to mirror the industrial liaison units of universities. Even where such offices have been established, their growth has been hindered, and in some case reversed, by the decentralisation of research functions described above. In their absence, there is disparity in the way in which research and technology is bought. The interplay between R&D or production units, finance and legal offices is likely to be vital. At worst, it is possible to imagine a situation where the inward licensing of new technology effectively requires the approval of several different departments Some areas, such as procurement departments or legal offices, may take a defensive approach to negotiations, others may be more concerned with the possibilities of the technology, without due regard for costs. Inevitably, some mechanisms will be needed to reconcile these differences, but in the absence of thorough research we cannot know how far they represent a barrier to the process. A project currently under way at the UK National Institute of Economic and Social Research, looking at determinants of intellectual property strategy in UK companies, might provide further evidence [12].

3. Institutions in the Technology Transfer Process

Studies of individual motivations must recognise that these can be discrete from those of their employing organisations. In some cases, individuals will be motivated to link their own activities to perceived expectations of their employers. It may be that the closer their employment to the centre of their organisation—such as the Industrial Liaison Officer in the central administration of his university—or the stronger the link between organisational performance and personal reward, the stronger this overlap of interest will be. In general, however, the UK authorities have devoted considerable resources to promoting the benefits of closer liaison, innovation and technology transfer to both companies and universities in recent years, without considering the extent to which it is in the interests of recipient organisations to cooperate. These attempts are often justified as an education process, facilitating better communication in the expectation that once the parties are aware of the opportunities available they will respond positively. But even with perfect knowledge, some companies and universities may conclude that greater attention to such matters is not in their interest. It is possible that the national interest, as perceived by government, will not coincide with that of the individual organisation.

The problems faced by government in influencing corporate behaviour are discussed in the section on policy implications. Even in the relatively stable western economies, there has been a need to stimulate and reward private sector innovation; Lederman's chapter describes some of the measures employed to that end. In the case of the United Kingdom, our conference heard officials from the Department of Trade and Industry describe how policy has shifted from support for the development of specific products and technologies to one of 'creating the environment' in which innovation would thrive. Measures to emphasise the value of successful innovation, and demonstrate best practice in the area included a high profile annual Innovation Lecture, publication of an 'R&D Scorecard' to highlight the companies devoting the highest proportion of their resources to such activities, the creation of a nationwide infrastructure of advisory services for business, particularly smaller companies, and of regional centres to promote specific programmes. Schemes to support technology audits in universities, and provide seedcorn funding for the expansion of university industrial liaison units, can also be seen under the heading of 'influencing the climate'.

The response of business to these initiatives has been mixed. It is still argued that both companies and venture capitalists take an uduly short-term view of financial gains. At the level of the firm, many appear unconvinced of the relationship between innovation and profitability. Reservations concerning specific R&D programmes from government, as expressed in reviews of the ALVEY, LINK and CASE initiatives, refer to their unashamedly pre-competitive nature—at a time when industry would value more highly assistance with the development of specific products—excessive bureaucracy and long lead times. Similar arguments are cited for non-participation, particularly among SMEs in European collaborative programmes.

At first sight, the benefits of better exploitation for universities appear obvious. Universities produce vast quantities of research each year, most of which is externally funded. As a result of legislation in the 1980s, they have greater ownership and exploitation rights than ever before, a trend which can also be seen in several west European countries [8] and the United States following the Boyh-Dole Act of 1980. Revenue generated from such exploitation—albeit by licensing to external companies—potentially represents a stream of revenue untied to any external commitment, at a time when the desire for new sources of income has never been greater. Failure to exploit this opportunity has typically been seen as reflecting either a lack of will or lack of information on the part of key actors.

On closer examination, there are several objections to this argument. It is becoming recognised that the financial benefits of technology transfer are easily overstated. Although UK universities are not required to publish their income from this source as a separate item, it is clear that this is both small and concentrated on a minute number of inventions. A recent survey by the Department of Trade and Industry [5] suggests that 80 universities showed an increase of £29.3 million from licensing their patents in the three years to 1994—an average of £122,000 per annum for each university, before any costs are taken into account. Even this modest income was very unevenly concentrated—over 90 per cent accruing to 21 per cent of the institutions sampled. Evidence from the US also suggests that significant returns have been confined to a small group of larger and, for the most part, internationally known institutions. Even here returns through licensing, although increasing, have rarely reached 5 per cent of research expenditure [1]. Moreover the costs, even allowing for much of the initial research being largely externally financed, can be prohibitive. It has been calculated that a medium sized UK university would require licensing income of approaching £5 million, at 1992 prices, over a ten-year period to 'break even' on investment in a significant technology transfer operation [9].

Without such investment the prospects for effective transfer are much diminished. Not only will effective marketing and legal skills be lacking, but even the basic protection of intellectual property may be overlooked. A report by the UK Office of Science and Technology [13] noted that:

> 'The general approach [of higher education institutions] is to commit as little of their own financial resources as necessary to file an initial patent application, then to use the following 12 month period to identify potential licensees who will assume the burden of subsequent patent costs in developing the invention further. There is a risk that this may lead to over-hasty and ill-considered judgements on long-term exploitability. Many inventions coming out of basic research need further development before their potential for exploitation becomes clear.'

Size of institution—or, more accurately, size of available portfolio of products with exploitation potential—is a critical factor in determining viability. Figures from the British Technology Group (BTG), set up by the UK government after World War II to assist with commercial exploitation of university based ideas, show that, of about 10,000 inventions investigated between 1949–91, only about 8 per cent generated any revenue at all, 4 per cent generated over £10,000 and just ten (0.1 per cent) produced a return of over £10 million [10]. It might be argued that individual universities, being closer to inventors, could enjoy a higher rate of success than BTG, but the figures presented are broadly in line with those for patents generally and, as the AUTM statistics show, their relevance is not confined to the UK alone.

Such figures may influence the willingness of universities to invest in the infrastructure necessary for effective technology licensing. Despite their increasingly entrepeneurial outlook in recent years, universities cannot be expected to adopt a long-term, high-risk investment strategy necessary without external assistance. Some would even question whether it is within their remit to do so. Many institutions still regard their core functions as being the provision of high quality teaching and research, and point out that technology transfer does not come under either heading. Unless clear incentives are available to encourage a different view, the support for technology transfer activities will continue to vary.

4. The Nature of Barriers to Technology Transfer

The chapters which follow offer perspectives on these developments from several different countries, and discuss the obstacles faced in each case. The nature, and relative importance, of these barriers can be expected to vary according to national circumstances. Most obviously authors from the transition economies of central and eastern Europe, where the process is being developed from a very low base, describe problems of a scale almost unimaginable to western practitioners. But the categories of problem, or types of barrier faced, show some consistency. Before looking at national situations in more detail, it may be sensible to list these.

4.1 LEGAL BARRIERS

Legal systems can be used to promote technology transfer, but can also hinder it. The experience of several countries shows the need for systems which give the confidence and incentives to individuals to protect their ideas, preferably without undue risk, cost or complexity.

In some cases, legal structures (or their absence) can influence technology transfer at a basic level. Some of the central and eastern European systems represented at our conference, for example, had yet to finalise the legal definition of corporate status; in others, the process of privatisation has been painfully slow. Even some western European states appear to have unduly complex procedures for maintaining corporate status, as noted in a recent European Commission White Paper on Innovation. In another case, excessive legal control over the process of international licensing had a restrictive effect on the process. A specific example, quoted by Pak and Türkcan, is the old Ottoman Patent law in Turkey, which prevented the payment of royalties for medical formulae on humanitarian grounds.

Until the 1980s technology transfer between universities and industry in several western states was hampered by legal structures which restricted collaboration opportunities. The extent to which universities could exploit their inventions was restricted by law in both the US and the UK until this period, and in the UK the British Technology Group held a right of first refusal on the majority of university owned inventions until 1985. In other countries the degree of collaboration with industry was itself restricted. Legislation to permit such activity, or clarify the terms under which it takes place, includes Spanish laws of 1983, 1984 and 1986, the Italian legislation of 1980 which provided a framework for academics to be rewarded for externally funded work, and French reforms of 1969 and 1984, which facilitated an industrial presence in the teaching and management of institutions, and provided a framework for staff secondment [8]. In addition to providing new opportunities, such legislation was seen as important in creating the climate in which further change could take place, as the Greek government put it, to 'create a new environment' for collaboration.

Permissive measures of this nature will only have maximum effect if individuals have the confidence to utilise them. A key element in ensuring such confidence is clear structures for the protection of new ideas once made public. For the former Soviet republics in particular this is no formality. In Egorov's paper on the Ukraine, for example, we hear how, despite the beneficial effects of legislation in the early 1990s, the process of gaining protection remains complex and difficult to access.

Once in place, the mechanisms for protection need to be simple and well understood. This theme united both eastern and western contributors to the conference, although the scale of problem they faced varied. In Romania, Sandu describes a situation in which parliamentary approval was needed for literally hundreds of laws within a short time scale, leading to delays—of at least four years duration in the case of the Law of Research—and an unstable legislative framework. We hear, for example, of approximately 38 new acts to regulate the activity of small and medium enterprises per month in 1994, building on 112 such acts issued in 1991–2.

This example goes beyond anything experienced in the west, but the importance of simplicity in legislation and licensing procedures is not confined to former Soviet states. Although the issues involved are wider than the narrow legal frameworks, the chapter by Hanlon and Gardiner, which describes difficulties faced by the small technology producer in dealing with large organisations in North America, provides a reminder that governments in all states need to ensure that regulations and procedures are as user-friendly as possible.

4.2 FINANCIAL BARRIERS

It is almost impossible to conceive of a world in which the availability of finance was not, in some way, a barrier to effective technology transfer. As with legal constraints, however, the problems faced by former socialist states of central and eastern Europe were of a quite different magnitude from those of the west. The accounts of Sandu, Todorov, Egorov and others all demonstrate not only that availability of finance is a barrier, but also the importance of ensuring that key actors are sufficiently entrepreneurial to seek it.

Given the quite overwhelming nature of the problems described, it is perhaps worth reinforcing two points. First, there will be financial constraints in any system. In our conference visit to the UK Department of Trade and Industry we heard of UK concerns that a generally profitable venture capital industry had failed to respond to the needs of new high technology projects, and particularly to investment opportunities in the range of, say, up to £250,000. The introduction of Venture Capital Trusts in 1995 represented one response to this problem.

Second, given that the majority of new technology will be both instigated and developed within industry, it is worth remembering that here too new ideas will have to compete for scarce funds. Nor will this process of competition depend entirely on the technological potential of an idea, as the following account demonstrates:

> 'It's important to recognise, and sometimes not appreciated at the academic–industry divide, that research and development budgets in industry are always limited; in fact, as the owner of a development budget, it is always limited and always overspent. So someone coming along one week before the final budget decision with a new idea, thinking that somehow I could find £10,000, £100,000 or whatever doesn't understand, doesn't appreciate the way in which industry funding operates. Every company will have its own financial model as to how much of its annual cash flow it is going to invest in research and development, marketing and production, and so on. And the amount that is spent in research will not be determined by how many good ideas they have got, rather the opposite, the good ideas they want to invest in will be determined by how much money they have got to spend' [15].

At first sight, neither of these points will offer much consolation to commentators from the transition economies. However, they do lead us to suggest that eastern and western countries can adopt at least two common approaches to their problems. First, since government cannot hope to resolve the availability of resources problem in the short term, it might do better to concentrate its efforts on the creation of an appropriate infrastructure. In this context it is interesting to note that the conclusions reached by Davies in Lithuania bear some resemblance to the strategy of the UK Department of Trade and Industry, which in recent years has moved away from direct financing of large-scale technology development projects, and concentrated instead on the creation of a 'climate' in which innovation can flourish, and on accessible, one-stop advisory networks.

Second, the question of how those public resources which are available for technology support should be targeted is an area of concern to both east and west. This concerns both the area of technology to be prioritised and the type of recipient most likely to benefit. In the former area, approximately half of the European Union states now conduct some form of 'technology foresight' exercise to assist with research priorities, and our conference saw examples of how similar procedures were being adopted in Poland and Romania, together with a wider analysis of civil R&D policies from Lederman. The need to analyse and evaluate initiatives of this nature will be a common problem for most of the countries represented at our conference over the coming decade. Our papers do, however, provide clear evidence of the need to take local needs and supply capacity into account in determining priorities; considerable duplication will result if such exercises concentrate too heavily on anticipating global trends.

Many governments also seek to target assistance by type of recipient. Although many UK and European Union schemes prioritise small and medium sized companies, the

TABLE 1. Resource problems

Resource problems	Key reasons
Financing	
Shortage of core funding	Sharp cuts in state funding; lack of alternative sources of funding; defaults
Working capital problem	Inflation; delays in payment
Supply of materials and equipment	
Physical and technological wearing out of equipment	Financial problems, general recession, tradition of keeping equipment in service for a long time
Lack of key components	Shortage of money, rupture of ties with old suppliers and uncertain ties with new suppliers
Poor quality of sub-components and ancillary parts	Tradition of underestimating the quality factor; lack of knowledge about world quality standards; very high prices for high quality sub-components
Need to obtain new super-precise devices	Transition to civilian mass production, increasing competition
Manpower resources	
The 'technical obsolescence' of personnel:	
Ageing of the labour force	Outflow of workers of the most productive age cohorts, poor recruitment of young people
Inadequate labour productivity	Low wages, loss of status of R&D and production work
Deterioration in human capital	Recession, part-time employment, de-skilling
'Physical outflow' of the best cadres:	
Domestic outflow to other spheres	Low wages in the sphere of S&T, absence of visible career prospects, buoyant Russian labour market in the business sphere
External 'brain drain'	Comparatively low standard of living, political and social instability, demand for Russian specialists in western countries

Source: Bzhilianskaya, L. [3].

definition of such enterprises has narrowed in recent years, both in terms of the maximum number of employees—now typically regarded as 250 rather than 500—and the relative merits of giving support for start-up companies, as opposed to those which have already generated a certain level of viability and are now seeking to expand.

Targeting of this nature will also be critical for eastern Europe, where the official number of small enterprises is huge. Todorov reports no fewer than 360,000 SMEs in Bulgaria alone.

4.3 MANPOWER CONSTRAINTS

Given the emphasis on the role of individuals described above, the quality and skills of those involved in the process will be vital. As Davies reminds us, the personnel necessary include not only scientists and technologists, but those capable of performing business advisory and support functions, those with R&D management skills, negotiation and legal expertise. Even in western states some of these professions are not well established. Although some attention has been paid to research management in recent years, it is still underdeveloped, particularly in the university sector.

Once again, however, these problems are minor compared with those reported from the transition economies, and it is not surprising that their most direct concern was with technical and scientific staff. From Romania, Sandu reports severe difficulties in attracting graduates in R&D functions—a 10 per cent decline in numbers below the age of 30 between 1991–3, and a further, though smaller, decline in the 30–39 age range. In a case study of the Russian Defence enterprises Bzhilianskaya reports similar problems, arguing that the problems of manpower are closely connected to those of finance, and are mutually reinforcing [3]. Particular concern is expressed for the outflow of people at their most productive age (estimated to be 30–45). The general problems faced in the area are summarised in the table above.

4.4 BARRIERS TO COMMUNICATION

Overall levels of contact between the science base and industry have increased significantly in the past fifteen years. In the UK, the volume of research grants to universities between 1982–94 rose in cash terms by a factor of five, a figure supplemented by greater industrial participation in government sponsored collaborative schemes. Over the same period, the proportion of university income from external research grants as a whole rose from 14 per cent to 21 per cent. The increased importance of industrial collaboration can also be seen in other western European states and the US. This is manifested not only in the growth of individual collaborators on specific issues, but the widespread adoption of structures which promote contact, for example the growth of science parks which place university and industry in close proximity.

Yet there is evidence that increases in the total volume of contact are not necessarily reflected in the volume of technology transfer activity. We have already noted that growth in collaborative R&D has outpaced that in the licensing of technology from universities to industry and there is ample evidence that the formal location of companies on university-based science parks does not necessarily lead to increased cooperation. Parker warns of the danger of adopting 'fashionable models' of interaction, without proper attention to the past lessons and a clear understanding of objectives.

Rahm highlights the need to define carefully the circumstances in which university–industry collaboration does not realise its full potential. These and other accounts lead us to consider whether level of contact and levels of effective communication may be quite different. It is possible that too many channels can exist. Small companies in the UK, with limited time and resources, have been faced with a bewildering choice of government grants, a range of competing partnership opportunities—including Regional Technology Centres, University Enterprise Trading Partnerships, Training and Enterprise Councils and other regionally-based schemes, not to mention the advisory services offered by the private sector. The possibility of confusion was a major reason for the establishment of locally based 'Business Links', which aimed to coordinate such activities under one roof.

Several speakers at our conference referred to cultural differences. Some of these will have a direct effect on the success of collaboration—for example, if the partners have different perceptions of the importance of keeping to budget or time deadlines, or if one partner is motivated by the desire to publish, whilst the other places emphasis on secrecy in order to maximise commercial advantage. Others may be less tangible, such as the feeling that industrial work is 'dirty' or that academics 'do not live in the real world'.

In the long term, such factors will best be addressed by the provision of incentives for individuals and organisations to work together, and structures which ensure that individuals in universities and industry meet together on a regular basis. This building of mutual confidence and respect will doubtless be a gradual one. But a review of current practice might reveal more immediate possibilities. Much could be done, for example, to improve the means by which university-generated technology is marketed to potential partners. Analysis of promotional material produced by universities suggests that the emphasis is still placed on developments which will enhance the general reputation of the institution rather than maximise the benefit to any single firm; it may be that too often recipients of 'technology offers' are left to work out the precise benefits for themselves. Also there may be a tendency for producers of new technology to over-stress the long-term benefits of their work; by contrast potential recipients often have neither the need nor the budget for revolutionary new technology—their need is more likely to be for incremental change which addresses specific problems.

Finally, effective communication requires a mutual understanding of the structure of both universities and individual companies. As the section on Participants in the technology Transfer above shows, both are complex organisations, and it cannot be assumed that the individuals within them share uniform attitudes and objectives. In such circumstances the existence of appropriate 'access points', with the time, inclination and sensitivity to examine new ideas, is vital. This is particularly so since many key actors regard technology transfer as being at best a part-time activity.

4.5 TECHNOLOGICAL BARRIERS

It may be that technology produced by science-based research organisations is not transferred to industrial partners because, quite simply, it is the wrong technology for their needs. We have already noted that universities have different missions, and are judged by different criteria, from commercial partners. In the section on communication, we speculated that this might lead to emphasis being placed on longer-term benefits of university inventions, leaving potential partners to work out for themselves the incremental benefits which could accrue in the short term.

Some of these problems can be resolved by better marketing and communication, but most require planning at the national level. Most governments covered by our conference have been forced to consider the extent to which they wish to see their science base concentrate on basic, strategic, applied or near market research, but clear definitions have been hard to sustain, and communicate to practitioners. Often this has led to unclear signals. In the United Kingdom, from 1992 to 1995, clear tensions developed between the Innovation Unit of the Department of Trade and Industry, which was established to encourage maximum relevance of academic work to UK companies, and the Office of Science and Technology, which sought to combine this aim with the need to preserve excellence. Added to this were the conflicting interests of the Treasury, which has warned against the use of university funds to subsidise industrial users, and the many government departments—most notably Defence and Health—which themselves act as customers for university work. Finally, the desire of government to use academic research as a means of wealth creation has to be reconciled with European Community competition law, which limits the extent of subsidy to individual firms and sectors.

A clear strategy will also need to take into account questions of cost. There are some who would claim that the emphasis in many European Union states (and by the Commission itself) on promoting 'pre-competitive' research programmes to small and medium-sized companies was fundamentally misguided. SMEs, so the argument goes, have neither the time nor the resources to invest in high-risk work which, at best, will come to fruition in five to ten years time, particularly if the intermediate results have to be made public in the meantime. Again, the assumption that such work will be attractive too easily ignores the incremental nature of innovation and the interests of key actors, both referred to above.

Finally, it may be that the 'best' or newest technology—which 'leading edge' university departments will consider it their mission to pursue—is not the most relevant to the needs of individual companies, or even national economies. This point is made, with particular reference to eastern Europe, by Davies in his study of Lithuanian agriculture.

5. Conclusion

The purpose of this introduction has been to demonstrate that the problem of technology transfer is more complex than often imagined, and is not likely to be resolved by any single measure, or package of measures from government. It has also been suggested that our analysis of the problem needs to extend beyond the macro level to those organisations and, most critically, those individuals who are expected to execute transfer in practice. Finally, we have noted that, although the scale of problem differs between countries represented at the conference, many of the basic issues are common to east and west. Each of these themes will be developed further in the more detailed chapters which follow.

References

[1] Association of University Technology Managers (US) (1993), Reports on '*Technology Commercialisation*', 3, 11.

[2] Betz, F. (1996), 'Industry/university centers for connecting industry to science', *Proceedings of the Unic International Conference on Technology Management*, Istambul.

[3] Bzhilianskaya, L. (1996), 'The transformation of technological capabilities in Russian defence enterprises, with special reference to dual-use technology', STEEP Discussion Paper no. 31, Science Policy Research Unit, University of Sussex.

[4] Corcoran, E. (1994), 'The changing role of US corporate research laboratories', *Research and Technology Management*, July–August.

[5] Department of Trade and Industry (1996), *Survey of Industry–University Research Links, 1995*, DTI Innovation Unit, London.

[6] Etzkowitz, H. (1996), 'Capitalising knowledge in a triple helix of academic–industry–government relations', Paper to Unic Technology Management Conference, Istambul.

[7] Feller, I. (forthcoming), 'Technology transfer from universities', in *Higher Education: Handbook of theory and Practice.*

[8] Kirkland, J. (1992), 'Cooperation between higher education and industry in the European Community. An overview', *European Journal of Education*, 27, 4.

[9] Kirkland, J. (1993), 'Financial technology transfer in UK universities', *Higher Education Quarterly*, 47, 1.

[10] Kirkland, J. (1994), 'Technology audit: from ivory tower to brass tacks', *IEE Review*, September.

[11] Mason, G. and Wagner, K. (1994), *High-level Skills and Industrial Competitiveness: Post Graduate Engineers and Scientists in Britain and Germany'*, NIESR, London.

[12] Matthews, D., Pickering, J., Kirkland, J. and Wilson, C. (1995–7), *The Determinants of Intellectual Property Strategy in UK Companies,* ESRC Research Project Ref. 325253023.

[13] Office of Science and Technology (1992), *Intellectual Property and the Public Sector Science Base',* HMSO, London.

[14] Perel, M. (1990), 'Discontinuities and challenges in the management of technology', *Research and Technology Management*, July–August.

[15] Robinson, G. (1994), 'Barriers to technology transfer', talk to ESRC Research Seminar Group, CRICT, Brunel University, April.

A COMPARATIVE ANALYSIS OF CIVILIAN TECHNOLOGY STRATEGIES IN FRANCE, FEDERAL REPUBLIC OF GERMANY, JAPAN, UNITED KINGDOM, AND THE UNITED STATES

L.L. LEDERMAN[1]
National Science Foundation
4201 Wilson Blvd., Arlington, Virginia, 22230, USA

This chapter compares the national civilian technology policies, strategies, and priorities of France, the Federal Republic of Germany (FRG) before reunification, Japan, the United Kingdom (UK), and the United States (US).[2] An analysis is given of similarities and differences; the national 'systems'; and data on R&D funding, science and engineering personnel and technological outputs are provided. The results of empirical research on US academic–industry cooperation and transfer are summarised.

1. Introduction

Technology and science are commonly accepted in the leading world trading countries as being major contributors to economic and social progress. The purpose of this chapter is to highlight the similarities and differences among four selected countries (France, FRG, Japan, UK), the European Community (EC) and the US. Each country has its own tradition of technology organisation and historical, political, and economic setting. They have used different means for determining national technological priorities and strategies and of allocating financial and human resources with different results. Each of these countries now faces a variety of pressures to modify its system in response to an increasingly competitive and interdependent environment.

The approach used in comparing the civilian technology policies and strategies is that of an objective analyst. This paper analyses what countries do rather than what they say; and avoids being either spokesman or apologist. Actions taken and decisions implemented, especially in the allocation of resources, are often more revealing of strategies than formal statements. Practice and actions are more important than preaching and can be used to analyse the policies, priorities and strategies of countries that have little by way of overt general policy statements as well as those that have formalised national policies and plans.

Each of the following sections could usefully be the subject of a lengthy paper in its own right. The space limitations here make it impossible to include details that are presented in other publications. References to such publications are provided for the reader who wishes to pursue a particular subject in greater depth.

J. Kirkland (ed.), Barriers to International Technology Transfer, 1–21.

2. Major Differences in Priorities and Structure

Several important overall differences between the US and the other four countries should be highlighted before looking at the individual countries.

There has been much discussion and some literature attempting to classify the Science and Technology (S&T) system in different countries. For present purposes, it may be enough to say that there is general agreement that the US R&D system and organisation are at the pluralistic, less centralised, and market-oriented end of the spectrum; the French system and organisation are at the more centralised, planned, and strategically targeted end of the spectrum; and the UK, FRG, and Japan are somewhere in between, depending on who is looking at what part of the system [20, 21].

These other countries each achieved a consensus, some time ago, that the central government has a clear responsibility to support S&T to serve civilian industrial needs. This includes supporting S&T to develop new and improved products, processes, and services, especially in areas of increasing international competition. This proactive policy is in part due to their smaller size, smaller domestic markets, constrained financial, natural and human resources, high proportion of GNP devoted to exports, and the aftermath of World War II.

In the US, no such consensus exists, and debate continues about such a strategy, with differing views in each of the industrial, political, and educational communities. Without such a consensus much of the US debate centres on organisational changes rather than on achieving a consensus on basic policies, priorities and mechanisms. There have been proposals to establish a Department of S&T (the Presidential Commission on Industrial Competitiveness), a Department of Industry and Technology, a Technology Foundation, a National Applied Science Administration, a National Civilian Technology Agency or Department and others. However, a reasonable case can be made that the absence of a more proactive US federal stance with regard to civilian commercial technology is, in itself, a policy decision, especially given the numerous executive and congressional reviews under both Republican and Democratic leadership during the past two decades.

By contrast each of the other countries has a specific ministry or department in the national government charged with the responsibility of furthering industrial S&T interests (see next section). The US government has no such organisation although the 1988 Omnibus Trade Act gives the Department of Commerce some additional responsibilities in this area. The current Congress is likely to reduce or eliminate this and may eliminate the Department of Commerce.

On 22 February 1993, President William Clinton and Vice President Albert Gore, Jr. issued a 36-page policy document and plan entitled, *Technology for America's Economic Growth, A New Direction to Build Economic Strength*. The document sets forth three goals:

- long-term growth that creates jobs and protects the environment via technology;
- making government more efficient and more responsive through technology; and
- world leadership in basic science, mathematics, and engineering.

The Clinton–Gore document includes the following key points:

(1) Develop a national network of manufacturing extension centres to help small and medium-sized businesses gain access to technology.

(2) Invest in applied R&D in fields such as advance manufacturing, aerospace, biotech-

nology, and advanced materials.

(3) Increase partnerships between industry and the national laboratories.

(4) Develop a partnership with the American auto industry to enable the development of a 'clean car', creating jobs and protecting the environment.

(5) Expand the Commerce Department's Advanced Technology Program to provide matching grants for industry-led R&D consortia.

(6) Develop a National Information Infrastructure and 'information superhighways' including:

(a) support for the Higher-Performance Computing and Communications Initiative that is developing new technologies for our most powerful computers—supercomputers that are able to process enormous quantities of information rapidly—and for a national high-speed network (information superhighways) to make this high-performance computing more accessible; and

(b) development of new applications for high-performance computing and networking in health care, lifelong learning, and manufacturing.

(7) Improve the environment for private-sector investment and innovation by:

(a) making the incremental research and experimentation tax credit permanent;

(b) reducing capital gains for long-term investments in small businesses;

(c) reforming our antitrust laws to permit joint production ventures.

(8) Greater government efficiency and responsiveness by:

(a) the federal government using technology to cut its costs, improve energy efficiency, and improve the quality and timeliness of service; and

(b) the government working with industry to develop technologies (software, computer, and communications equipment) that increase the productivity of learning in our schools, our homes, and our work places.

(9) Enhance the management of United States technology policy by:

(a) high-level leadership and coordination by the Vice President, the Office of Science and Technology Policy, and the National Economic Council;

(b) developing a true partnership among the federal government and industry, labour, academia, and the states; and

(c) regular evaluation of programmes to determine whether they should remain part of the national investment in technology. Recent Congressional changes indicate a less proactive policy and the likelihood of a rollback of some recent proactive initiatives.

In each of the other countries there is greater emphasis than in the US on focusing or strategically targeting national S&T efforts on areas believed to be important for future economic development. These areas include electronics, computers, informatics, biotechnology, materials, robotics, and manufacturing technology. Support in such areas for academic research and education, and industrial R&D and commercial activities is considered 'strategic', and increasingly government and private resources are being provided. Since each of these countries is targeting many of the same technological areas, overcapacity could result that would make recoupment of public and private investment difficult and could put further strains on international technological competition and trade [19, 20, 21, 22, 33].

Except for the US and Japan, the higher education of students in these countries is supported by the central government as a social overhead. All qualified students have a right to higher education—often in the field and institution of their choice if space is available—with low or no tuition costs and often with stipends. Recently such support has been reduced. In Japan, support is provided in the form of no interest loans repaid over a period of 13–20 years after graduation, graduate student fellowships, and graduate research awards. It is important to note, however, that a smaller proportion of the college-aged population in the European countries participates in higher education than in the US and Japan. Research and education are less often coupled in the other countries than in the US and FRG, with education having more of a pedagogical and less of a research orientation [19, 20, 22, 24]. Most important for transfer of knowledge, know-how and technology, the relationships between academia and industry have tended to be weaker in the other countries than in the US. Historic distrust and disinterest have existed, with academia feeling that research of industrial interest is less desirable, and industry feeling that academia could contribute little to its needs. Recently this distrust and disinterest have lessened, and numerous bridges are being built, in part as a result of government policies and incentives [20, 22, 24].

Historically, engineers have been accorded higher status in Japan, FRG, and France, than in the US or the UK. This is in part shown by a higher prestige, proportionally more degrees, and greater proportion of engineers in top industrial, academic and government positions [14]. For example, Japan, with one-half the population of the US, graduates about the same number of engineers.

The mobility of faculty members and industrial and government scientists and engineers is relatively low in the other countries as compared to the US. It is not unusual in these other countries for faculty members and R&D personnel in other sectors to spend all or most of their careers in one organisation. Greater effort and central government incentives have been applied recently to encourage more mobility between sectors and movement to strategically important S&T areas in order to improve knowledge, know-how, and technology transfer [20, 22, 33]

The other countries place much greater emphasis than the US does on formal international cooperation in S&T, both bilateral and multilateral. A considerably higher proportion of their central government funds supports such cooperation than in the US. A much higher proportion of their graduate students and postdoctoral researchers do their work in other countries (often in the US) and significantly greater effort is made to keep up with progress and literature from other countries [19, 20, 22].

Except for the US, the other countries have national university systems with government(s) providing general university funds (or GUF) plus extra support that may be needed for specific projects. Faculty members' salaries are paid for by government(s) and faculty are essentially lifetime civil servants.

In contrast with the US, new and additional efforts are being made to assess and/or evaluate the effectiveness and results of S&T efforts in these countries. For example, on 23 December 1985, the French government passed a law that requires the Ministry in charge of S&T 'to present a yearly assessment to Parliament of the "strategic choices" of national science and technology policy, presenting the position of France in international competition in comparison with major foreign countries.' In the UK, the S&T Assessment Office was established to help Departments, Research Councils and so on assess the results of their R&D expenditures. In addition, the Royal Society and the Royal Academy of Engineering in 1986 established the Science and Engineering Policy Unit (SEPSU) to provide assessments and advice. In the FRG, the Ministry for Education, Science, Research, and Technology has a technology assessment group that compares German S&T with that of other leading countries. In Japan, assessment of S&T progress is a continual part of the consensus mechanisms used to establish goals and objectives. This includes the large DELPHI study conducted periodically to assess the views of scientists and engineers from a broad range of government, university and industrial organisations. The National Institute for S&T Policy (NISTEP) is designed, in part, to perform such assessments.

3. Specific Country Information

The sections below provide a brief summary of the central government policies, organisation, strategies, and special programmes for civilian industrial technology in each of the other nations and the EU.

3.1. FRANCE

France operates the most centralised and planned system of these countries. The government provides a higher proportion (51 per cent) of total National S&T funding than in the other countries. The Ministry of Research and Technology was recently incorporated into a Ministry of National Education, Higher Education, Research and Social Insertion (implementation). It is responsible for the coherence of the National S&T policy and for the major part of federal R&D funding. Other important ministries in the field are Telecommunications and Space, and Industry. Public S&T is performed in agencies which carry out the government policies and priorities in scientific research (CNRS), space (CNES), atomic energy (CEA), technology transfer and innovation (ANVAR), health and medicine (INSERM), oceans and fisheries (IFREMER), and energy conservation and renewable energy (AFME).

A top priority is the promotion, development, and increase in industrial S&T. In order to improve the relatively low proportion of R&D financed by industry, the government has introduced a series of direct and indirect incentives. Specific national programmes include: aeronautics, telecommunications, biotechnology, production technology, electronics, new materials, nuclear and other energy technology and space [14, 25, 26].

3.2. FEDERAL REPUBLIC OF GERMANY

The government of the Federal Republic of Germany believes that in a free market economy the primary responsibility for civilian technology rests with industry and relies heavily on the private sector which is responsible for the majority of S&T funding (over 60 per cent). Nevertheless, certain areas, like Airbus, are heavily subsidised. A wide spectrum of industrial technologies benefit from government S&T funding, both directly and through various European programmes. Further strengthening of private sector initiatives are assisted by tax incentives as a part of the 1990 tax reforms, current venture capital pool, and government procurement.

The Ministry of Research and Technology (BMFT) is the main federal organisation, accounting for almost 70 per cent of the federal government's support for civilian R&D. About 40 per cent of the federal government's R&D expenditure is devoted to the category 'promotion of market-oriented technology.' Included in this category are nuclear and other energy sources, information processing, electronics, microelectronics, production engineering, materials, supersonic aviation, biotechnology and ground transportation. The Economics Ministry provides additional funding to promote market-oriented technology development. Recently BMFT has been combined with the Ministry of Education and Science into the Federal Ministry for Education, Science, Research and Technology (BMBF). Following a thorough review of the new Länder's academy institutes, most have been closed or converted and special funding has been concentrated on a few Max Planck Institutes, about ten Fraunhofer Institutes and several groups at academic institutions in the new Länder following reunification, that are funded by the Max Planck Gesellschaft for five years and then assessed to see if they should continue to be autonomous Max Planck Institutes [4, 15, 27].

3.3. JAPAN

Japan's most fundamental S&T strategy mechanism is the setting of national policy through an emerging consensus judgement. The most important formal coordinating body is the Council for Science and Technology in the Prime Minister's Office, composed of ministers, senior educators, industrial managers, scientists, and engineers. Special councils are formed periodically to assess progress in different fields and recommend priorities. Government agencies, including the Ministry for International Trade and Industry (MITI) and The Science and Technology Agency (STA), operate research institutes whose work is planned and conducted in close cooperation with industry. STA also performs important coordinating and advisory services in the national government. Each agency acts independently and STA and MITI employ an array of advisory councils and industry associations to ensure that government-conducted and government-sponsored research will be consonant with private sector S&T interests.

Japan's S&T has reached a turning point, according to the latest annual White Paper on research released in mid-July 1995. During the prolonged recession, both industrial and government R&D funding have declined for the first time since the 1950s. R&D as a percentage of GNP has also declined [35].

Noteworthy among STA supported programmes is Exploratory Research for Advanced Technologies (ERATO). It supports teams from industry, academia, and government that are led by key individuals in programmes of interdisciplinary breakthrough R&D. MITI has an elaborate system in support of industrial S&T, carried out in its own laboratories and through active promotion of privately supported research institutes. The focus is on performing non proprietary R&D in certain product areas (for example, semiconductors and new synthetics) and focused R&D initiatives. MITI's National Project System focuses R&D in national priority areas. The programmes are carried out primarily with industry participation and costs are shared by industry and government. Industry provides a higher proportion of total national S&T funding (about 70 per cent) in Japan than in any of the other countries. In fact, industry provides 98 per cent of the R&D performed in industry and the government only 2 per cent.

The government actively encourages private S&T by a system of financing and tax incentives. Included are favourable interest rates (for example, from the Japan Development Bank), contracts for commercialisation of innovations, capital investments, and conditional interest free loans. Tax incentives provide special tax credits for incremental R&D above previous levels, R&D performed by small and medium-size firms, and the cost of depreciable assets for R&D in basic technologies (for example, new materials, biotechnology, high performance robotics). Special programmes include: information technology, Computers, semiconductors, new synthetics, new materials, robotics, energy and resources, the 'human frontier', and the Key Technology Centres [1, 2, 16, 17, 29, 30, 38].

3.4. THE UK

In July 1995, the British Prime Minister announced the move of the Cabinet Office's Office of Science and Technology (OST) to the Department of Trade and Industry (DTI). The move was justifed on the grounds that it 'will allow the government's policy in science, engineering, and technology to be developed alongside its policies of industry, and with due regard to the contribution of science, engineering, and technology to long-term wealth creation'. OST will still be responsible for the several Research Councils and the recent Technology Foresight activity.

The DTI is the main government department concerned with industrial applied research and technology, but other organisations also have major functions. The LINK programme, which cuts across a number of Departments, funds up to one half the cost of developing technologies in areas such as molecular electronics, eukaryotic genetic engineering, nanotechnology, and biotransformations. Other specific programmes include: manufacturing and information technology (Alvey and its successors), opto-electronics (JOERS), advanced materials, biotechnology, aeronautics, micro-electronics and CAD-CAM [3, 5, 6, 7, 8, 11, 12, 13].

3.5 EUROPE

To complete this summary, mention should be made of some of the special European cooperative S&T efforts which are growing in scope and size and may become even more important. It should be pointed out that while EU cooperative S&T efforts are notable and expanding, they have so far been relatively small. One observer estimates that 'the entire budget of the European Community is less than that of, say two major ministries in just one of the 12 member states; and of this, R&D gets only 3.1 per cent' [18, p. 412].

Under the formerly approved 1994–8 EU Framework, the main research and technology programmes and funding for these five years (1 ECU = US$ 1.34) are: Information and Communications Technology ($1.5 billion), Industrial Technology ($2.7 billion), Environment ($1.4 billion), Life Sciences and Technology ($2.1 billion), Energy ($3.0 billion), Transport ($0.3 billion), and Finalised Socio-economic Research ($0.2 billion).

A number of the European countries are cooperating outside the EU framework. The most significant effort is the EUREKA programme which extends European basic and precompetitive S&T into areas 'closer to the marketing of technological products' [9, p. 67]. The activities are jointly supported by the West European governments and industry. One example project is the microelectronics cooperative programme, the Joint European Submicron Silicon Initiative (JESSI) which has attracted considerable support from the EC countries. EUREKA's total cost 1985–90 is about $4 billion, of which about 40 per cent has come from the French government and industry [9, 10, 23, 36, 37].

It should be noted that this paper focuses on industrial civilian technologies for the non-government market place. The EC and the various countries frequently have large nuclear, space, etc. technology programmes in addition to those discussed.

4. Data on National S&T Efforts and Outputs

One of the best ways of describing and comparing national S&T strategies, priorities and results is by examining data on financial (in OECD purchasing power parity) and human resources and outputs [32,34]. Ideally we want measures of S&T but comparable data on inputs exist only for R&D which is somewhat narrower and does not cover the follow-on technology investments (for example, engineering design and manufacturing) [28, 31].

TABLE 1. Indicators of the science and engineering effort relative to country size

Indicator	US	Japan	FRG	France	UK
GNP, 1989 (bns of constant 1982 $)	4118	1446	722	645	681
R&D, 1989 (bns of constant 1982 $)	111	46	21	15	13
R&D/GNP ratio	2.7%	3.0%	2.9%	2.3%	2.0%
Non-defence R&D/GNP ratio	1.9%	3.0%	2.8%	1.8%	1.6%
R&D scientists and engineers ('000s)	949	442	166	109	101
R&D scientists and engineers per 10,000 labour force	77	69	54	45	36
Non-academic engineers per 10,000 labour force	188	187	182	113	137

Note: Data are for a year in the 1987–9 period, depending upon country and item.

Table 1 shows indicators of the science and engineering effort relative to the size of each country. Total R&D as a per cent of gross national product is similar in magnitude ranging from 2.2 per cent for the UK to 3.0 per cent for Japan, 2.9 per cent for West Germany and 2.7 per cent for the US. For non-defence R&D as a percentage of GNP, the range is greater: from 1.6 per cent for the UK to 3.0 per cent for Japan (and 1.9 per cent for the US). In the US 60 per cent of government R&D funding is for defence; in Japan only 4.8 per cent. The current US administration has announced its intention to reach a 50 per cent defence and 50 per cent civilian balance over the next several years.

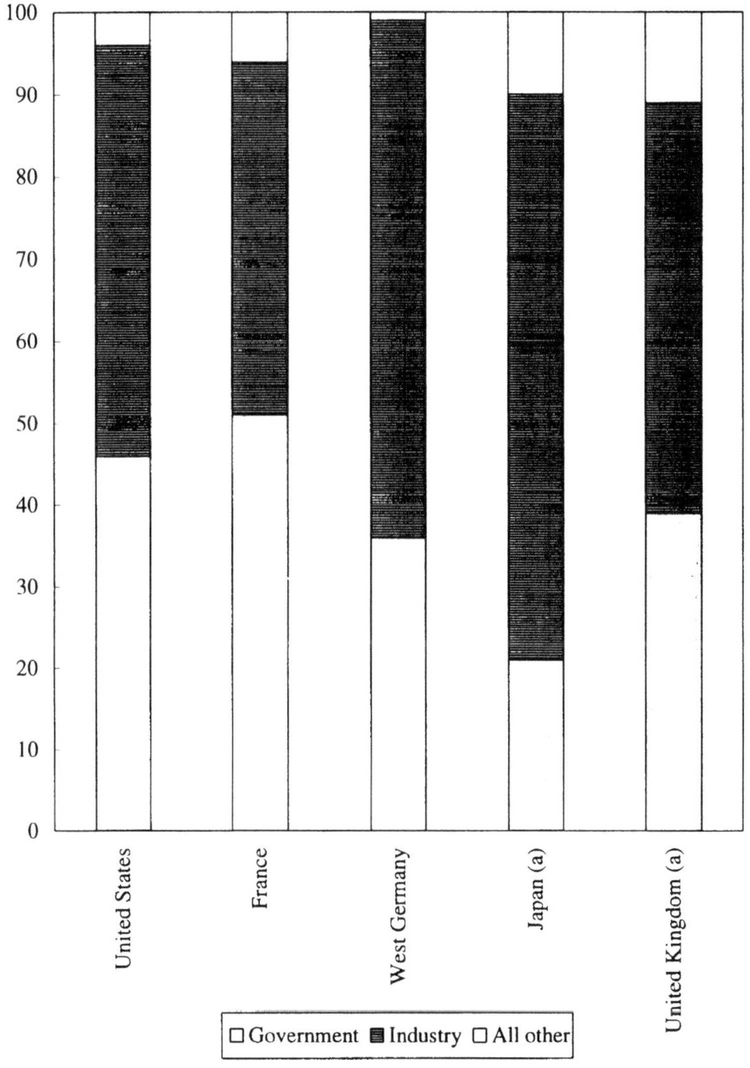

Figure 1 Distribution of R&D expenditures by source of funds: 1988
(a) Data are for 1987.

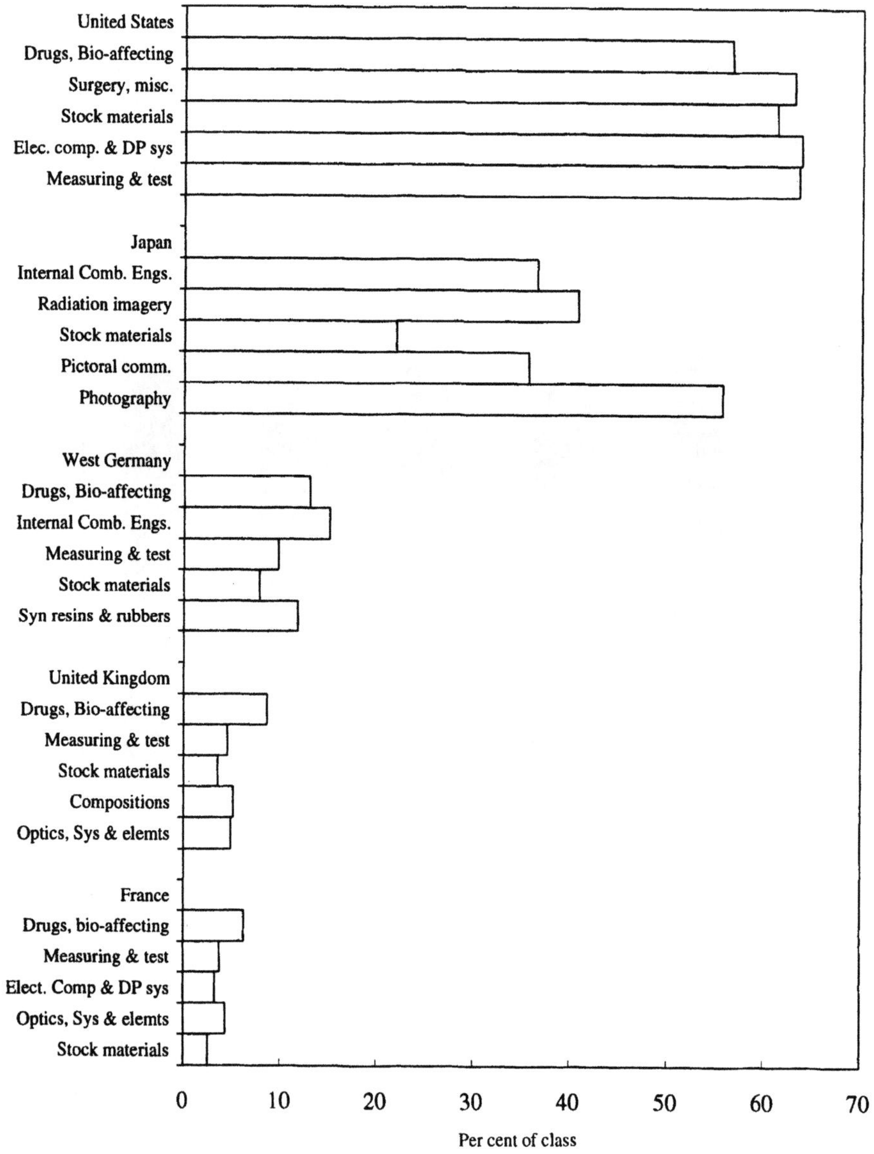

Figure 2 Shares of US patent classes in which inventors from selected countries have the largest numbers of patents: 1975–88

The US and Japan have the highest number of R&D scientists and engineers per 10,000 persons in the labour force, 77 and 69 respectively, with FRG, France and the UK at the low end of about 54 down to 36. For the absolute size of GNP, R&D funding, and the number of R&D scientists and engineers, the US is larger than the total of the other countries combined.

Figure 1 shows the distribution of R&D expenditure by the source of the funds. The relative contributions of government and industry to R&D funding and performance vary significantly by country, reflecting differences in their industrial structures, government policies, and patterns of government defence-related R&D spending. Japan and the FRG have the largest shares of industrial R&D funding (72 per cent and 65 per cent, respectively); France has the smallest (43 per cent) and the UK and the US next (51 per cent). Conversely, France has the largest share of government R&D funding: 51 per cent (the US is 47 per cent). The UK has a somewhat anomalous distribution of R&D funds by source, receiving 37 per cent from government, 51 per cent from industry and 11 per cent from other. This distribution largely reflects the relatively higher proportion of R&D funds the UK receives from abroad (Cabinet Office, series).

Figure 2 displays the shares of US patents in which inventors from these countries have received the largest number of patents in the 1975–88 period. Bio-affecting drugs is the patent class in which most FRG, French, and UK inventions are patented in the US. The largest number of Japanese-origin US patents are in the area of internal combustion engines. Japanese inventors received 37 per cent of all US patents in this class and 50 per cent of all US patents in photography. US inventors have large numbers of patents in the stock materials, electronic computers and data processing systems, surgery, measuring and testing and bio-affecting drugs patent classes. US-held patents in each of these categories represent at least one-half of all such patents.

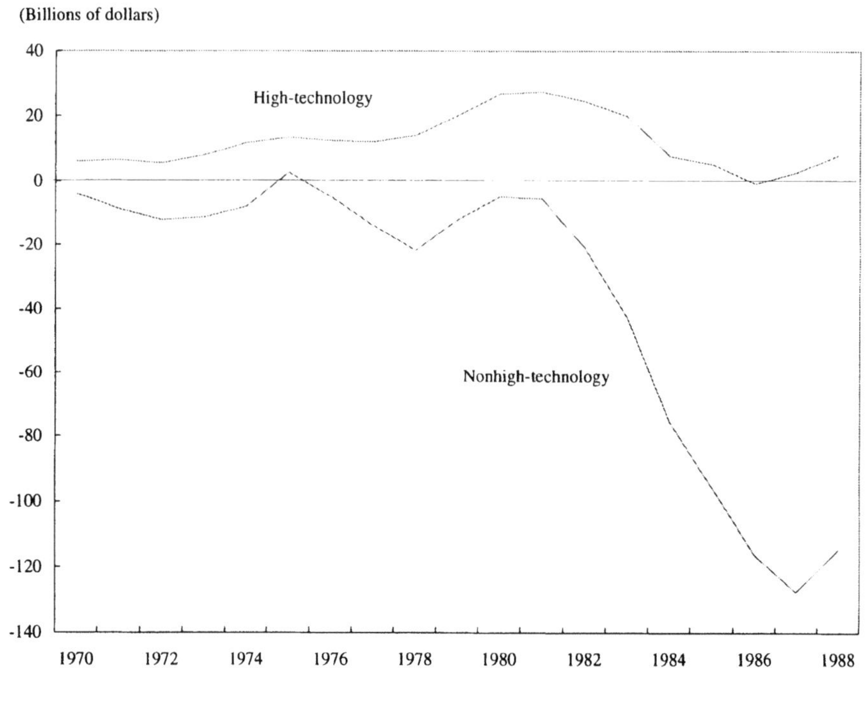

Figure 3 US trade balance in high-technology and non-high-technology manufactured products

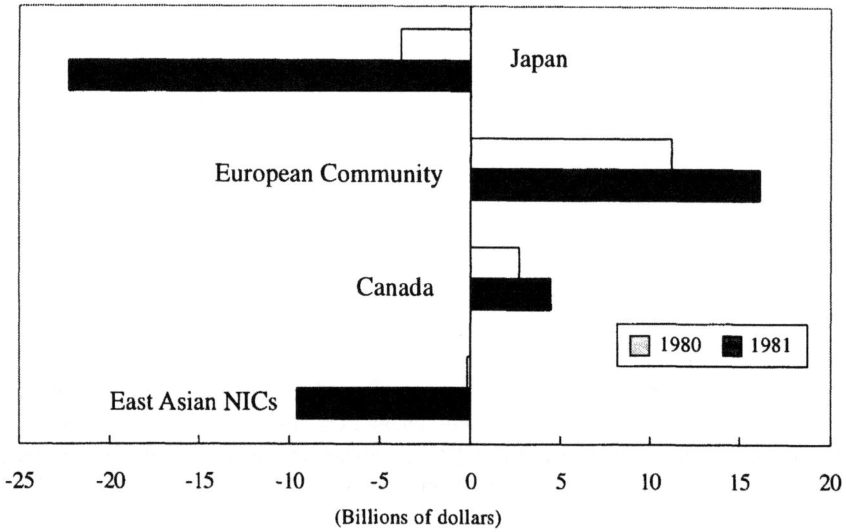

Figure 4 US trade balance in high-technology manufactured by selected countries and regions, 1980 and 1988

Figure 3 shows what has happened to the US trade balance in high technology and non high-technology manufactured products from 1970 to 1988. High technology here is defined as the R&D intensity of supplier intermediate and capital goods industry as well as final products. The US had a high-technology trade balance of $8.1 billion in 1988 and $2.7 billion in 1987 after experiencing its first deficit in this category in 1986. By comparison, the US had a trade surplus of $26.7 billion in high-technology products in 1980. High-technology goods accounted for 41 per cent of US exports of manufactured products in 1987, up from 34 per cent in both 1980 and 1974.

Data on the US high technology balance of trade indicates that the US has done reasonably well in exports of such products but has imported much more from other countries into the US [28]. Balance of trade in high technology by region indicates that between 1980 and 1987 the US balance with the European Community and Canada was positive and growing more positive; with Japan and the East Asian newly industrialised countries it is negative and growing more negative (Figure 4). In 1988, Japanese products accounted for 35 per cent of all US high-technology imports, and the $22 billion US high-technology trade deficit with Japan is almost six times that of 1980. The US high-technology trade balance with Europe and Canada has increased from 1980 to 1988.

Figure 5 displays 1965–87 world export shares of technology intensive products from 24 reporting countries on exports to and imports from, each of nearly 200 partner countries. Technology-intensive products are defined as those for which R&D exceeds 2.36 per cent of value added. Japan is the only one of these five countries which has steadily increased its share of technology-intensive exports from 1965 to 1984. The Japanese share of such world exports nearly doubled between 1970 and 1987; in 1987, it was about equal to that of the US—19 per cent versus 21 per cent. After recovering from a low of 21 per

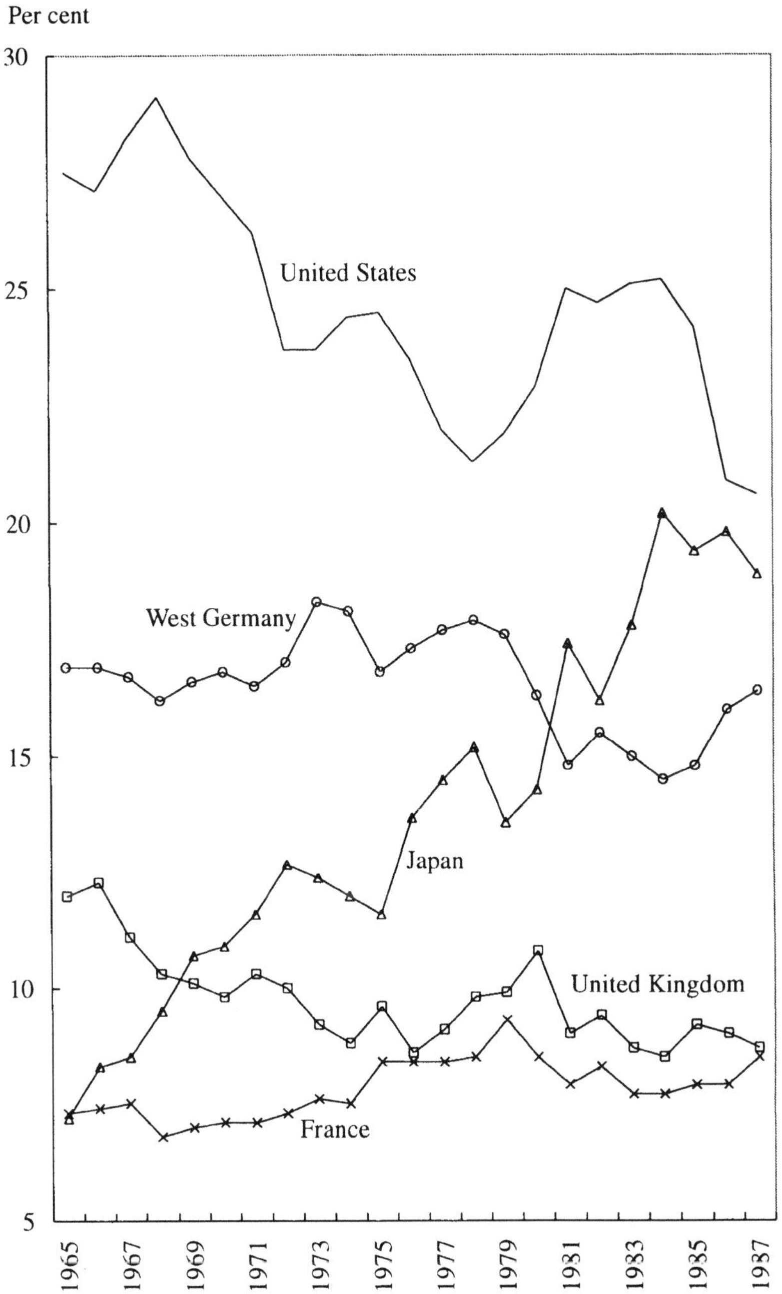

Figure 5 World export shares of technology-intensive products

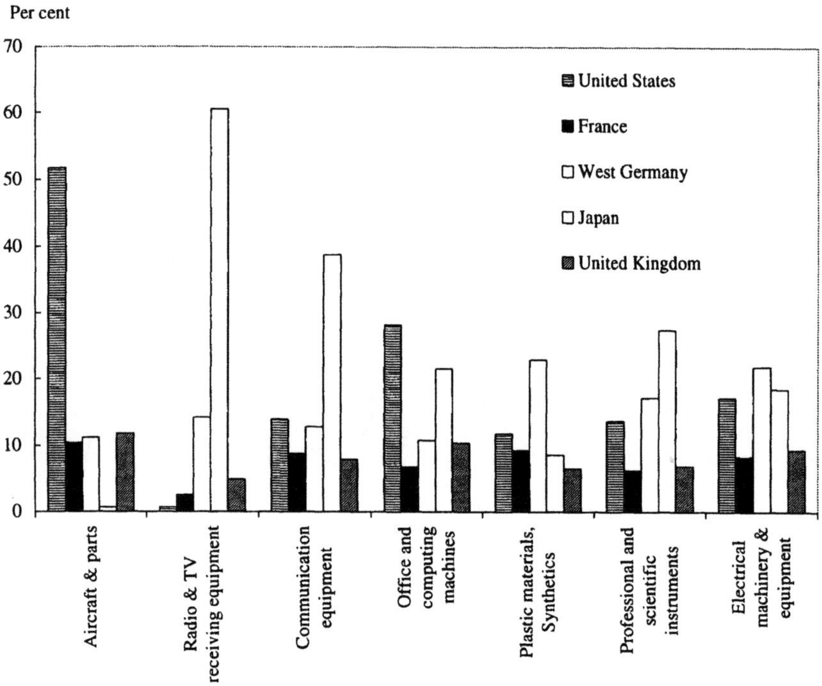

Figure 6 World export shares of technology-intensive products by selected product field, 1987

cent in 1978, the US world export share of technology-intensive products remained stable between 24 per cent and 25 per cent from 1981 to 1985. However, the US share declined significantly from 1985 to 1986, dropping from 24 to 21 per cent; this was largely because of increased exports from countries other than the four discussed here.

Figure 6 shows world export shares of technology-intensive products by selected product field in 1987. The US dominates in aircraft and parts and leads in office and computing machines. West Germany leads in elastic materials and synthetics and electrical machinery and equipment. In addition, the US leads in agricultural chemicals and engines and turbines (not shown in Figure 6) and equipment. Japan dominates in radio and TV equipment and communications equipment and also leads in professional and scientific instruments.

5. Options

Before concluding we should ask what options this cross-national comparison suggests. However, before addressing this question, the following cautions should be noted. The elements of a system for supporting and conducting S&T are interrelated with each other and with the broader national context. Individual elements should fit into the overall national context and are not necessarily transferable from one country to another. While each country is looking at the policies and strategies of other countries with a view to what can be learned and possibly applied, the other countries are doing the same. A number of changes have been introduced that could move the countries closer together in their policies and strategies. However, their organisation and resource allocations remain very different. Finally, the advantages and disadvantages of optional policies and strategies are not clear. There is little objective assessment information revealing what works better or worse under what circumstances and why. The positive and negative consequences of a particular option frequently depend on how it is implemented.

With these cautions in mind, it is nonetheless useful to consider some options in light of changing objectives and needs. This can be useful even if it serves the purpose of reinforcing commitment to current policies and strategies, with or without some modifications. The following list of questions are offered for consideration based upon the 'Major Differences in Priorities and Structure' discussed earlier.

1. First and foremost, should the US move further towards greater government support for nonproprietary S&T of use to industry, especially in areas of increasing international technological competitiveness? Each of the other countries accepts this as a part of the responsibility of the central government and has employed both direct and indirect mechanisms to discharge it. In the US no such consensus exists, and debate continues about the advantages and disadvantages of such a strategy, with differences of view in each of the political, academic, and industrial communities. Proposals by the current Administration would move further in this direction, but without congressional consensus.

2. Is it desirable to focus graduate support on fields where demand is growing strongly and is greater than supply? For the US, there is also the question of whether it should shift towards more direct support for graduate students. Most US S/E graduate students receiving financial assistance from the federal government are supported as a part of project grants to faculty; most graduate students in the European countries are supported by low or no tuition costs, fellowships, and often stipends.

3. Should the US engage in more cooperative activities with other countries, especially where the costs of facilities and equipment are high? As discussed above, each of the other countries allocates proportionately more resources to such cooperative ventures than does the US. With regard to international cooperation, there are some indications that participation by US scientists and engineers may not be as high as it was in earlier periods. The advantages and disadvantages of the various forms of cooperation (for example, bilateral, multilateral, sharing of decision-making) in particular fields of science and technology need to be carefully considered.

4. Should the US move toward greater centralisation or coordination of government S&T activities and needs? This question has frequently been raised in the US. Of course, centralisation is not synonymous with coordination or quality, and no good evidence can be drawn from the experiences of the countries examined to support the greater efficacy of more centralised versus more pluralistic systems. These countries run the gamut in this regard, and several countries have shifted along the spectrum in both directions.

6. Academic–Industry Research Cooperation and Transfer

This section will summarise the results of recent empirical research using US academic and industry respondents at various levels of their cooperation. This research deals with: (1) what is and is not going on in the US; (2) how are potential conflicts avoided or dealt with; (3) what are the attitudes of those cooperating and those not cooperating; and (4) what are the inputs, outputs, impacts and payoffs of academic R&D? This general subject has been controversal with different presumptions, attitudes, and ideas. The results of these studies, involving broad-based and representative groups of academic and industrial staff, raise questions about some of these beliefs.

The following highlights a few key points from the five empirical studies supported and organised by the National Science Foundation's Research on Science and Technology programme, directed by the author:

From Rahm's study of 'University–Firm Linkages for Industrial Innovation' we learned that:

- the initiation of university–industry interaction largely happens through informal mechanisms, such as faculty consulting and contacts from alumni.
- the extent of university–industry interaction can be enlarged by the creation of formal friendly mechanisms, such as research parks, cooperative agreements, class offerings, and professional refresher courses.
- firms which cooperate with universities are quite proactive and identify university researchers of interest,
- barriers to university–industry linkages stem largely from different cultures and values and can be extensive by centering around negotiation of intellectual property rights and lack of incentives within universities to promote faculty involvement with firms.

From the Albert Rubenstein and Eliezer Geisler study of 'Relations Between University–Industry Interaction and Industrial Innovation', we know that:

- the development of long-term relationships is multi-faceted and often includes consulting, support of students, participation in university research centres, and research contracts,
- although there are many grants and gifts from firms, most contracts are between operating units of the firm and individuals, groups, or centres in a university,
- younger faculty are not trained, or in many cases, even encouraged to be entrepreneurial or proactive with firms,
- there is little incentive or sanction for the managers of operating units to encourage their technical people to interact with university researchers, especially given their short time horizons,
- When attempting to assess the impact of transfers from universities, firms have difficulty in tracing and imputing the source.

Slaughter and Campbell's study of 'Protecting the Public's Trust: A Search for Balance Among Benefits and Conflicts in University–Industry Relationships' found that:

- collaborative activity is supported as long as it does not interfere with teaching responsibilities,
- inappropriate activities are attributed to individuals, not organisations,
- respondents do not acknowledge possible internal equity issues related to collaborative activity and institutional incentives to stray from appropriate conduct.

Yong Lee's study of 'The Academic Research Climate and Technological Innovation', which focused on changes in climate and attitude toward university–industry relations found:

- academics in the 1990s report that they are more likely than in the 1980s to grant research credits for patentable inventions in connection with tenure and promotion and to accept user oriented research as having a legitimate place in university research.
- while academics offer strong support for their university to participate actively in economic development efforts and work closely with industry to commercialise academic research, they express serious reservations about their university crossing the traditional academic boundary by providing startup financial assistance to new technology-based firms or owning equity in firms based on university research,
- academics today are under great pressure to seek outside funding for research; this coupled with their sense of responsibility, appears to influence their transfer-supportive attitudes.
- faculty support for university–industry cooperation is tempered by fear, real or imagined, that close cooperation might interfere with academic freedom, basic research, and academic integrity.

Edwin Mansfield, in a series of studies of academic research and industrial innovation, found:

- high economic returns to society (20–30 per cent a year) from university R&D in the form of industrial innovation and about 10 per cent of new or improved products, processes, and services were based upon academic research in major science-based industries, both in 1975–85 and in 1986–94,
- in 1986–94, the time interval between academic research findings and the first commercial introduction seems to be significantly shorter than in 1975–85, which enhances the economic payoff to society from academic research,
- a positive, but not high, correlation between the geographic proximity of the university and the firm that benefited, with important exceptions,
- practically all of the academic researchers cited by industry as having contributed significantly to product and process innovations received substantial support from the federal government, and their percentage of support from industry is about twice that of other academics,
- second tier academic departments (based on National Academy of Sciences ratings), not just the prestigious ones, contributed significantly to industrial innovations.

A gestalt summary of key collective findings includes:

- the transfer of knowledge, know-how and people from universities to industry is far more important than the transfer of technology, which firms regard as their business for the most part.

- The processes by which such transfer occurs are multiple and complex. What is clear is that transfer is very much a 'body contact sport' and that in this arena, publishing and circulating papers is far less important than direct conversation.
- The most important contacts are those between one or more university and industrial researchers as a part of their professional interests and institutional responsibilities, and especially as consultants.
- Non-researchers, such as transfer agents or patent and licensing people, can be important as facilitators of the transfer process; however, they should be regarded as necessary, but not sufficient. The researchers themselves are the most important part of the transfer process, and facilitators should be very introspective about when not to get involved, and when to reduce promotional activities and stay in the background.
- Most research universities and their cooperative firms have figured out mechanisms that allow them to pursue their mutual and separate interests, while avoiding potential conflicts and criticisms.
- Most universities and industry researchers do not participate in cooperative efforts and that's fine because each contributes to the interests of the other in less direct ways and should probably be left alone. The relatively few that are cooperating have been shown to be of disproportionate importance, and the payoffs are very high. They should be encouraged in every sensible way.

Finally, what we do know about mutually satisfying industry–university cooperation has been significantly increased by studies such as these, but it is still limited. Research, experimentation, testing, and evaluation are needed to determine what works better and worse, under what circumstances, and why.

As for international technology transfer, it should be noted that the major vehicle for such transfers is the transfer between industrial organisations. This includes firms and their foreign subsidiaries, partners, or other permanently or temporarily related industrial organisations.

7. Conclusion

This chapter demonstrates that there are many similarities and differences in the S&T policies, strategies, priorities, and practices among these countries—especially differences between the US and other countries. In analysing the comparative advantages and disadvantages it appears as if some of the advantages when carried to an extreme can become liabilities. For example, the US reliance on pluralistic, shorter-term project support contributes to greater flexibility, mobility, and market-orientation. It also results in less stability, less proportionate investment in infrastructure, and less general support for graduate students than exists in other countries. Such advantages and disadvantages tend to be reversed in most of the other countries which tend to provide longer-term programmatic support principally from a single agency or ministry.

It is not surprising that some of the more important recent changes appear to be designed to lessen the disadvantages while maintaining the advantages. In this way the countries can learn from each other, and adopt or adapt that which serves their individual country needs, while seeking to preserve their unique advantages.

Notes

1 The author is Senior Staff Associate and Director, Research on Science and Technology Program, Directorate for Social, Behavioral and Economic Sciences, National Science Foundation, Washington, D.C. 20550.

2 This chapter is based upon an ongoing assessment of the science and technology policies, priorities and programmes of the US and other countries. The author gratefully acknowledges the assistance of staff members at NSF and other US government agencies, the Science and Technology Counsellors of foreign embassies in Washington, D.C., US Science and Technology Counsellors overseas, officials in the ministries of science, technology, industry and education responsible for S&T, staff in the OECD and EC, and individuals in academic and industrial S&T in each of the countries. A special acknowledgment goes to Jennifer Bond, Programme Director, Science and Engineering Indicators, Division of Science Resources Studies, NSF. The analysis presented in this article is that of the author and should not be interpreted as reflecting the views of NSF, the US government or the individuals who provided information, review or comments.

References

[1] Agency of Industrial Science and Technology (1988), AIST, September 1991, Ministry of International Trade and Industry, Agency of Industrial Science and Technology, Tokyo.

[2] Anderson, A.M. (1984), *Science and Technology in Japan*, Longman Group, London.

[3] 'British science sent back to the dog-house' (1995), *Nature,* 376, 13 July, 101.

[4] Bundesministerium fur Forschung and Technologie (1988), Statistische Informationen, February, Bonn, Author.

[5] Cabinet Office (series), *Annual Review of Government Funded R&D*, HMSO, London.

[6] Cabinet Office Advisory Council for Applied Research and Development (1986), *Exploitable Areas of Science*, HMSO, London.

[7] Central Office of Information (1988), 'Research and Development in the United Kingdom in 1986', *Economic Trends*, August, London, Author, pp. 82–8.

[8] Central Office of Information (1989), *The Promotion of Science and Technology in Britain*, London, Author.

[9] Commission of the European Communities (1987), *The European Community of Research and Technology,* 18, Luxembourg, Author.

[10] Commission of the European Communities (1988), *The First Report on the State of Science and Technology in Europe*, Luxembourg, Author.

[11] Department of Trade and Industry (1988), *DTI—the department for enterprise,* January, White Paper, HMSO, London. See especially Chapter 8, 'Innovation, using technology', pp. 33–7.

[12] Dickson (1995a), 'Concern and anger greet shift of UK science unit into industry agency', *Nature,* 376**,** 13 July, 103.

[13] Dickson (1995b), 'Major claims policy change will "strengthen" UK science', *Nature,* 376, 20 July, 206.

[14] Embassy of France (1989), 'French Advances in Science and Technology', Washington D.C, Science and Technology Office, Embassy of France. Newsletter; see especially Summer, 1989 issue.

[15] Federal Ministry for Research and Technology (1988, abridged version), 'Report of the Federal Government on Research', Bonn, Author.

[16] Gamota, G. and Frieman, W. (1988), *Gaining Ground: Japan's Strides in Science and Technology*, Cambridge, Mass., Ballinger. Covers: computers, opto- and microelectronics, advanced polymers, mechatronics, telecommunications, and biotechnology.

[17] Institute for Future Technology (1988), *Future Technology in Japan: Forecast to the Year 2015*, Tokyo, Author.

[18] International Science Policy Foundation (1988), 'Science and technology in Europe', *Science and Public Policy*, December, London, Beech Tree Publishers, 366–430.

[19] Lederman, L.L. (1985), 'Science and technology in Europe: a survey', *Science and Public Policy,* June, 131–143. Covers: France, Federal Republic of Germany, the Netherlands, Sweden, and United Kingdom.

[20] Lederman, L.L. (1989a), 'Science and technology policies and priorities: a comparative analysis', *Science*, 4 September 1989, 1125–33. Covers: France, The Federal Republic of Germany, Japan, Sweden, the United Kingdom, and the United States.

[21] Lederman, L.L. (1989b), 'US research and development policy and priorities and comparisons with selected countries: Canada, France, The Federal Republic of Germany, Japan, Sweden, the United Kingdom and the United States', in Abu-Laban, B. (ed.), *University Research and the Future of Canada*, University of Ottawa Press, Ottawa, pp. 475–505.

[22] Lederman, L.L., Lehming, R., and Bond, J. (1986), 'Research policies and strategies in six countries: a comparative analysis', *Science and Public Policy,* April, pp. 67–76. Covers: France, Federal Republic of Germany, Japan, Sweden, the United Kingdom and the United States.

[23] McLoughlin, G.J. (1989), *The Europe 1992 Plan: Science and Technology Issues*, March 16, Washington, DC, Library of Congress, Congressional Research Service.

[24] Martin, B. and Irvine, J. (1989), *Research Foresight, Creating the Future*, London, Pinter Publishers. Covers basic and strategic research; France, FRG, US, UK, Japan, Canada, Australia, Sweden, and Norway.

[25] Ministere de la Recherche et de la Technologie (1986), *Recherche et Developpement Dans Les Entreprises*, Paris, Author.

[26] Ministere de la Recherche et de la Technologie (1988), *Rapport Arnexe sur L'Etat de la Recherche et du Developpement Technologique: Activities en 1987 et 1988 et Perspectives*, Paris, Author

[27] National Science Foundation (1986), *The Science and Technology Resources of West Germany: A Comparison with the United States*, Washington, DC, National Science Foundation, Division of Science Resources Studies. NSF 86–310.

[28] National Science Board (1988),*Science and Engineering Indicators —1987*, Wash-

ington, DC, National Science Foundation. Biennial publication.

[29] National Science Foundation (1988b), *The Science and Technology Resources of Japan: A Comparison with the United States*, National Science Foundation, Division of Science Resources Studies, Washington, D.C. NSF 88–318.

[30] National Science Foundation (1991), *International Science and Technology Data Update*, NSF, Division of Science Resources Studies, Washington, DC. Published annually.

[31] National Technical Information Service (1989), *Directory of Japanese Technical Resources in the U.S*, Springfield, VA, National Technical Information Service, PB 89–158869.

[32] Organisation of Economic Cooperation and Development (1986), *Science and Technology Indicators—Resources Devoted to R&D*, Paris, Author. Biennial publication.

[33] Organisation of Economic Cooperation and Development (1988a), *Science and Technology Policy Outlook*, 1985, Paris, Author.

[34] Organisation of Economic Cooperation and Development (1988), *Main Science and Technology Indicators 1982–88*, Paris, Author. Science and Technology Indicators—Basic Statistical Series.

[35] Science and Technology Agency (1991), *White Paper on Science and Technology 1991*, September, Tokyo, Ministry of International Trade and Industry, Science and Technology Agency.

[36] 'Science in Europe', *Science*, 4 September 1987, pp. 1125–1188.

[37] 'Science in Europe', *Nature*, 27 April 1989, pp. 717–736.

[38] Swinbanks, D. (1995), 'Japan tackles downturn in science funding', *Nature*, 376, 20 July, 207.

TECHNOLOGY TRANSFER IN PRACTICE

PROMOTING TECHNOLOGY TRANSFER IN THE US UNIVERSITY:
When it Works, When it Doesn't

DIANNE RAHM[1]
Department of Political Science
515 Ross Hall, Iowa State University, Ames, Iowa 50011–1204

1. Introduction

Technology transfer-oriented university–industry interactions take many forms. They range from short-term specific activities with clear 'deliverables' to on-going interorganisational associations with relatively non-specific outcome expectations. The numerous types of technology transfer-oriented university–industry interaction are best understood from an interorganisational viewpoint. The connections firms make with universities and the sort of associations they engage in are determined in large part by the company's strategy and particular characteristics. At the same time, research universities are complex organisations comprised of many semi-independent centres, research units, offices, academic departments, and laboratories under the shared umbrella of university policies and procedures. The connections made between universities and firms vary in form and type based both upon the university sub-unit that is the focus of interaction and a specific university's policies and procedures. Of course, the most fundamental 'sub-unit' in the university structure is the individual professor.

US universities and researchers are motivated by a host of factors to create and maintain linkages to the business community. US Federal government policy has provided a financial incentive by permitting universities to generate revenue streams by licensing technologies. Policy guidelines from granting agencies, recommending that researchers demonstrate the social relevance of their research, have also had an impact. Federal and state grant programmes with the eligibility requirement of joint university–firm proposals have created incentives for partnering. The service mission of universities provides yet another stimulus. Making new technologies generally available is viewed by universities as a public service. This sense of fiduciary responsibility is an especially strong inducement for state universities. Part of their service mission is state-based job creation, through the university spinning-off new technology-based start-ups, and state-based job retention, through university technical assistance to the state's companies. Uncertainty about future streams of federal funding in the post-Cold War era has had an impact. Universities and university researchers are adapting to their new environment, in part, by seeking industrial money as a possible substitute source of funds.

The business community can take advantage of much that universities have to offer. For example, small start-up companies may license their 'enabling technology' from the

J. Kirkland (ed.), Barriers to International Technology Transfer, 25–41.

university. Well established firms might look to the university for product line growth and productivity improvements. Firms with substantial R&D capability might view ties to university researchers as enlivening and enriching their own internal R&D effort. Firms experiencing fiscal pressures and downsizing of their R&D efforts might look to university research as a replacement resource. The outcome expectations of firms and forms of interaction with universities are diverse. The cornucopia includes (but is not limited to) businesses employing university researchers as private consultants, affiliating with university-based research centres, sponsoring university research, licensing technologies, hiring graduates, using university equipment and library facilities, contracting research or services from university-based units, entering into free materials or equipment sharing agreements with university researchers, endowing chairs, or supporting research conferences.

Since the interactions between universities and industry are dominated by interorganisational complexity, the question of 'what works, and what doesn't' cannot be asked generically. Instead, the query needs to be addressed in the context of the variety of outcome expectations and the myriad forms of interaction.There is no one answer to this question. Successful university interchange for a large firm experiencing internal R&D downsizing will likely be very different than what would be viewed as successful by a new start-up dependent on the university for its enabling technology or by a rapidly growing high-tech company with its own strong internal R&D component. From the university perspective, one must distinguish factors influencing the individual researcher's decision to undertake technology transfer from technology transfer efforts engaged at the university level. What works, and what doesn't work will shift with the unit of analysis.

Drawing on data gathered both from written questionnaire and personal interviews, this chapter explores the nature of technology transfer between US universities and industry. After reviewing some of the literature, the question of 'what works and what doesn't' will be discussed in the context of interorganisational complexity, the variety of outcome expectations, and the forms of interaction. Suggestions for areas of further research are offered.

2. What Works and What Doesn't: Some Background Information

Much of the effort that has gone into studying university–firm collaborations has concentrated on the kinds of interorganisational interaction. One way to understand this interaction is to look at the development of university bureaucratic structures with the responsibility of handling industrial liaison [2]. The presence of new organisational structures indicates an investment on the part of the university and a permanence of the link with industry [10]. Such bureaucratic structures might include offices of technology transfer, technology licensing, industrial relations, business incubators, industrial extension services, and research parks [7]. It is generally thought that centre affiliation, teamwork, and an interdisciplinary focus [3] will enhance the level of university–industry interaction, thus the presence of centres on campus are a visible sign of this interaction [9][11]. Universities may also engage in activities aimed at generating contact with firm personnel, such as sponsoring university technology expositions [6], or offering courses at times and on subjects of interest to firm personnel [12].

These forms of interaction tend to be formalised in some university structure but they exist alongside a host of informal university–firm interactions. Some of these informal interactions include former students (now employed in a firm) calling a prior professor to seek answers about a firm project, university researchers engaging in paid industrial consulting, or firm personnel directly contacting researchers regarding their expertise.

The literature suggests that creating internal university procedures to encourage researcher participation in industrial efforts is also important because the social and organisational environments of universities and firms are often in conflict. The university's tradition of holding in high value (and thus rewarding) publication of basic research results in many ways conflicts with the needs of industrial collaborators [4].

While universities have many reasons for involvement in technology transfer related activities, the potential financial gain from such activities raises special concerns. The desire to generate a revenue flow from the commercialisation of technologies has far more important implications for the university as an institution than the simple development of free-standing bureaucratic offices to manage industrial relations [1][5]. With this activity comes the need to establish policies within the university dealing with the appropriate balance between applied and basic research, internal university rewards for researchers who participate in technology transfer-oriented interactions, and the inevitable conflicts of interest that arise when researchers and universities collaborate with firms [8].

The literature is helpful in outlining some primary factors of interest when studying university–industry technology transfer-oriented interactions. Still, the question of what works and what doesn't is not explicitly answered.

3. Methods

This chapter draws on results coming from both a written questionnaire and a series of personal interviews. The written survey was sent to researchers in departments of Biology, Chemistry, Computer Science, Electrical Engineering, and Physics at the top 100 US research universities. The initial population list of the top 100 universities was drawn from National Science Foundation data where universities were ranked based upon annual R&D expenditure. These universities include a high percentage of land-grant colleges and are geographically well distributed across the nation. The top 100 universities account for the vast majority of R&D expenditure at all US universities. The disciplines included are each ranked highly as recipients of R&D funds and research results coming from these areas are likely to be directly related to industry. In total, 1,774 participating researchers were sent surveys in September of 1993. Of those, 1,013 completed researcher surveys were returned yielding response rate of 57 per cent.

The second phase of the study involved site visits to selected universities and firms. Five universities were selected for site visits. Their selection was determined in large part by an attempt to ensure geographic diversity and to mix public with private universities. Another important criterion for site selection was the willingness of the contact person to work with the author in arranging interviews with university and business personnel. Site visits were completed between August and October of 1994 at The Pennsylvania State University, Georgia Institute of Technology, the University of Wisconsin—Madison, Stanford University, and Washington University. In total, 70 interviews were conducted with university and industry personnel (48 personal interviews with university personnel and 22 personal interviews were conducted with high-level managers and scientists in firms). For the purpose of confidentiality, the locations and names of those interviewed will not be revealed.

4. What Works and What Doesn't: The University Viewpoint

4.1 QUESTIONNAIRE FINDINGS

One way to determine 'what works' from the university perspective is to draw on the mailed questionnaire results to create contrasting profiles of university researchers—those who engage in technology transfer interactions with firms (hereafter referred to as the *industry-linked researchers*) and those who do not (hereafter referred to as *university-bound researchers*). Several questions of interest can then be answered. What are the characteristics of industry-linked researchers that differentiate them from university-bound researchers?[2] What factors best account for the individual researcher's decision to undertake technology transfer related activities? What factors explain the extent of technology transfer related efforts on-going at any one particular university?

Industry-linked researchers differ from their university-bound colleagues in the number and type of interactions with firms. As Table 1 shows, industry-linked researchers can be characterised as having extensive *personal* interactions with companies as a result of their own aggressive outreach to companies, industry outreach to them, engaging in paid

TABLE 1. Differences by personal contacts

	Industry-linked	University-bound	Chi-square
Researcher	(663)	(91)	267.0****
contacts firm	87%	36%	
Firm contacts	(717)	(135)	247.9****
researcher	95%	53%	
Tech. manager	(343)	(57)	41.93****
introduction	45%	22%	
Researcher	(564)	(67)	189.4****
consults	75%	26%	
Alumni contact	(608)	(126)	100.2****
researcher	80%	51%	

(Numbers in parenthesis are observed frequencies)
**** = p-value < .0001

TABLE 2. Differences by department offerings

	Industry-linked	University-bound	Chi-square
Student	(288)	(72)	7.65***
internships	40%	28%	
Professional	(249)	(41)	25.86****
workshops	33%	16%	
Satellite	(171)	(23)	22.31****
courses	23%	9%	
Courses taught	(164)	(24)	18.61****
at firm site	22%	9%	
Night classes	(209)	(42)	12.35****
	28%	17%	

(Numbers in parenthesis are observed frequencies.)
***= p-value < .001
****= p-value < .0001

TABLE 3. Differences by university organisations

	Industry-linked	University-bound	Chi-square
Research	(474)	(139)	4.75**
consortia	62%	55%	
Industrial	(248)	(58)	8.74***
extension	33%	23%	
Joint R&D with	(534)	(123)	40.15****
firms	70%	48%	
Spin-off	(315)	(68)	17.56****
enterprises	42%	27%	
Research parks	(309)	(85)	4.20*
	41%	33%	
Industry-funded	(342)	(83)	11.98***
facilities	45%	33%	
Incubators	(237)	(57)	7.12**
	31%	22%	
Personnel	(217)	(40)	16.57****
sharing	29%	16%]	

(Numbers in parenthesis are observed frequencies.)
* = p-value < .05
** = p-value < .01
*** = p-value < .001
**** = p-value < .0001

industrial consulting, and continuing contact from former students at work in the business community.[3] While formal introductions to companies through university technology managers happen with greater frequency for industry-linked researchers than they do for university-bound researchers, this form of contact with firms is clearly not as important as a personal connection.

There is much universities do to foster interaction with firms besides having university technology managers make introductions.Many universities place emphasis on establishing firm-friendly class offerings as well as university-level organisations and activities. Do these firm-friendly department offerings and firm-friendly university-wide activities work? To explore this, the survey asked researchers a series of questions regarding department offerings that might be considered supportive of industrial linkage such as: graduate courses regularly scheduled in the evening (to facilitate access to firm personnel), student internships, professional workshops or short refresher courses, satellite courses broadcast to firm sites, and classes held at firms sites.Table 2 reports these results.[4]

The respondents were also asked to comment on whether or not their university has or sponsors any firm-friendly organisations or activities including: research consortia, technology transfer conferences, industrial extension services, cooperative R&D agreements with firms, technology expositions, research parks, industry supported university facilities, start-up firm incubators, or personnel sharing between the university and firms. As Table 3 shows, statistically significant differences exist between industry-linked researchers and university-bound researchers on all but two of these variables.[5]

As the last two tables show, industry-linked researchers present a different profile from university-bound researchers in regard to the number of them that seem to come

TABLE 4. Differences by research organisation and attitude

	Industry-linked	University-bound	Chi-square
Patents are not	(331)	(71)	18.27****
rewarded	44%	33%	
Feel pressured to	(417)	(122)	1.09**
do T-square	55%	48%	
Applied research	(308)	(134)	6.56****
hurts basic research	41%	54%	
Work primarily as	(154)	(70)	5.83**
sole PI	20%	28%	
Inter- or multi-	(584)	(157)	27.6****
disciplinary	77%	62%	
Affiliated with a	(456)	(113)	23.17**
centre	60%	44%	

(Numbers in parenthesis are observed frequencies.)
** = p-value < .01
**** = p-value < .0001

TABLE 5. Decision to engage in technology transfer
(Logistic Regression Results n=974)

Dependent variable: researchers's choice to engage in technology transfer activities Independent variables:	Statistical significance
Index of formal firm-friendly department and university offerings (a)	NSS
Researcher works primarily as sole PI	NSS
Researcher is interdisciplinary	NSS
Researcher is affiliated with a centre	NSS
Researcher takes initiative to contact firms	****
Researcher has been contacted by a firm	****
Researcher has been introduced to firm personnel by university technology manager	NSS
Alumni contact researcher	***
Researcher holds a patent	**
Researcher consults for industry	****
Researcher feels pressured to adopt applied research focus	***
Researcher thinks tech. transfer hurts basic research mission of university	NSS

NSS = Not Statistically Significant
**= prob < .01
***= prob < .001
****= prob < .0001

(a)This is a simple additive index taking values of 0 to 16. The presence of any one of the following would increment the index by one: student internships, professional workshops, satellite courses, courses taught at firm sites, weekend classes, night classes, research consortia, tech transfer conferences, industrial extension services, cooperative agreements with firms, spin-off enterprises, R&D expos, research parks, industry funded university facilities, business incubators, and personnel sharing with firms.

from universities with firm-friendly departmental or university offerings. The profile of industry-linked researchers also contains another aspect of university organisation. Of prime consideration are interdisciplinary research, team research, and centre affiliation.

The organisation of the research effort can be described along the dimension of traditional departmental, single disciplinary, primary investigator focus and the extent to which the researcher departs from this locus. As Table 4 shows, industry-linked researchers tend to be team-oriented, interdisciplinary, and centre-affiliated. Another aspect of the profile that differentiates industry-linked researchers is that they are more likely to hold patents than their university-bound colleagues[6] as well as to believe that the university does *not* adequately reward efforts resulting in patents. Industry-linked researchers are less likely to believe that the emphasis on industrial outreach is having a negative effect on the basic research mission of the university.[7]

TABLE 6. Extent of technology transfer
(Regression Results n=759)
R-Square: .25

Dependent variable: Number of firms interacted with in an attempt to transfer knowledge, know-how, or technology Independent variables:	Parameter	Statistical significance
Index of formal firm-friendly department and university offerings	0.40	****
Researcher works primarily as sole PI	-0.92	NSS
Researcher is interdisciplinary	-0.15	NSS
Researcher is affiliated to a centre	0.81	**
Researcher takes initiative to contact firms	2.20	***
Researcher has been contacted by a firm	0.63	NSS
Researcher introduced to firm personnel by university technology manager	0.86	NSS
Alumni contact researcher	2.10	****
Researcher holds a patent	0.21	****
Researcher consults for industry	2.03	****
Researcher feels pressured to adopt applied research focus	-0.58	NSS
Applied research does not hurt basic research mission	0.95	*

NSS = Not Statistically Significant
*= prob < .05
**= prob < .01
***= prob < .001
****= prob < .0001

The descriptive summary offered above is useful for creating a profile of industry-linked researchers but a question remains regarding each factor's relative importance. To explore this, two models were run. The first is a model of choice (logistic regression) used to better understand what factors distinguish researchers who decide to engage in technology transfer (that is, the industry-linked researchers) from their university-bound colleagues. The dependent variable under consideration is the 'yes or no' choice either to participate with industry or not to participate. The second model attempts to explore factors that influence the extent of technology transfer activities on campus. The dependent variable, the extent of technology transfer activity, is represented by the number of firms the researcher reports interacting with. A standard regression analysis was run. The results of both models are presented in Tables 5 and 6.

Table 5 highlights the importance of informal communication and personal connections. Contact with former students now working for firms, engaging in paid industrial consulting, and being personally sought out by firms because of their expertise each help to explain the profile of a researcher who engages in technology transfer activities. Two other variables have statistical significance in this model. Holding a patent distinguishes technology transfer engaged researchers. Finally, these researchers feel pressured to engage in technology transfer activities because they believe that university administration and government granting agencies will look favourably upon such activity.

In Table 6 the results of the analysis for explanators of the extent of technology transfer involvement reveal several differences from the model of choice. The formal communication channels and linkage mechanisms are highlighted. The greater the number of firm-friendly department and university offerings, the greater the *extent* of technology transfer firm interaction reported. Also of importance is the social and organisational environment. Centre affiliation on the part of researchers tends to increase the extent of technology transfer engagement. Informal communication and personal contacts are also explanators of the extent of technology transfer interaction. The personal initiative exercised on the part of the individual to personally approach firms is significant. Note that holding patents, being contacted by former students now in firms, and engaging in paid industrial consulting are significant in this model as well as in the first model presented.

4.2 INTERVIEW FINDINGS

The interviews with faculty and university staff provided a deeper understanding of technology transfer-related issues from the university perspective. The following discussion summarises the main points gathered from the series of interviews with university faculty and staff.

All the universities studied, whether public or private, land-grant or not, share a common sense of fiduciary responsibility. When the university is fulfilling this public service mission, universities tend to think their technology transfer-oriented interactions with industry are working. Universities have multiple measures to assess this activity. One primary indicator is the count of the number of technologies the university is able to help the business community commercialise (as new or improved products). Universities are particularly interested in creating jobs through the formation of new companies or the expansion of existing companies. Placement of students in firms where they can put new technology and information to work is of great importance, especially in terms of providing companies with the next generation of the captains of industry. Implementation of change within a firm, particularly by assisting firms to meet regulatory requirements through improved technology, is part of the fiduciary mission of universities.

The organisational mission of universities, to create and pass on knowledge, can act to foster industry linkages. Publication of research results in the scholarly journals, as well as presentation of papers at professional conferences, provide high visibility for university talent. Interested industry personnel reading the journals or attending the conferences can use these media to make personal contact with a particular researcher. The academic reputation of the researchers and the high standing of the universities attract interest from the business community.

Clear goals work. Careful planning that creates reasonable expectations on the part of all participants is important. Lack of detailed project plans (clearly stating the problem, the strategy, and who is providing what) can result in unreasonable expectations and failure. One way to implement this is when industry is active in setting the research agenda on joint projects.

From the university perspective, technology transfer-oriented interactions work when they can maximise university revenues from royalties and licensing fees without exposing the university to liabilities or financial risks.

Organisational structures, such as university multidisciplinary centres, work. Researchers can meet industry personnel there who may be interested in the researcher's work. Centre directors have the gate-keeping function of discovering what the companies need and understanding what the university can provide. Centres are seen as facilitators of industry–university interaction, but not simply because of their formal outreach efforts. Rather, centres are seen as filling a newly created gap. Science is becoming more sophisticated and expensive. Maintaining a lab is a costly affair. Centres can provide shared equipment resources and thus are more cost-effective. Because industry also uses the centre (because of the same financial pressures), academic researchers and business personnel are given the opportunity to meet face to face and establish a relationship. The existence of many contract research labs on campuses is also seen as a facilitator of university–industry collaboration. These labs, in the past, worked primarily for the government, but are flexible and can easily shift to fill an industrial need.

University distributed organisational structures as well as a general lack of uniform and well-understood policies often create problems. Universities that have centralisation of administrative function and distribution of authority in terms of dealing with the business community cite this organisational feature as a real facilitator. The structure of universities, however, is often not centralised. The business community prefers centralised authority while universities often spread decision-making across different offices and programmes. This can create confusion and a response time-lag that may be burdensome to firms. The rapid rise of formal offices of technology transfer and industrial liaison along with the increasing professionalism of university staff in dealing with industry, however, facilitate collaborations.

Negotiation of intellectual property rights often doesn't work well. For instance, when firms and university personnel collaborate, they are often required by university bureaucracies to work out intellectual property rights in advance, that is, before research begins or anyone knows what might result from the research. Sometimes this creates frustration. The central question, which takes considerable effort to come to an understanding about, is 'Who owns it?' Answering this question includes addressing other issues, such as companies arguing that if they are willing to pay the university's indirect costs they have a claim on any intellectual property resulting from a collaboration. This type of expectation on the part of companies is often seen as unreasonable by universities. Universities point to other unreasonable business expectations such as a demand for an exclusive licence for the life of the patent, with few diligence milestones built into the agreement.

Industry can be frustrated and impeded by the dominance of the academic calendar. Professors and research assistants can be brought on to a project usually only on a semester or quarter schedule. Industry, running on a twelve-month calendar, has some degree of difficulty adapting to this.

From the university perspective, interacting with firms can be difficult when firm requirements run counter to the perceived main mission of the university. This includes when firms are unwilling to include students in projects, or demands on the part of industry that faculty members delay or not publish results stemming from an R&D effort. Another example of this cited by those interviewed included the need for universities to hold on to what they are and fight the tendency to become an industrial service lab as some firms seek to out-source their R&D efforts and grant money available to universities declines. Engaging in socially relevant research is valued but the university must not become a 'service-shop' to industry and in doing so lose focus on its central mission—education and the creation of new knowledge.

The organisational culture of the university can be a major barrier to university–firm interactions. For instance, researchers would generally prefer to be funded by a traditional and well-respected source of academic funds (such as the National Science Foundation) as opposed to industry. Non-tenured researchers feel the need to 'prove themselves' as scholars first before engaging in applied activities. The culture issues also play out in the internal incentive systems of universities. In the university, patents are not treated as equivalent to publications nor is much value placed on industry-relevant work that results neither in patents nor publications (for example, a feasibility study).

5. What Works and What Doesn't: The Viewpoint of the Business Community

Company expectations of their university interactions vary widely. To better understand the interactions between universities and firms from the perspective of the business community, it is useful to create a typology of firms. Firms' expectations of university interactions vary with type. From the interview data, a typology of four firm types was created. What works and what doesn't work is best explored in each of these four specific contexts.

Type 1 companies are small high-tech spin-off companies. They are heavily dependent on university ties for the essentials of company survival: personnel, information flows, technology, and services. Very often the company's business centres around one 'enabling technology' that emerged from the university that the firm now licenses. The company founder may be a university professor or the firm may employ the inventor of the enabling technology on a permanent consulting basis. The university itself may be a client of the Type 1 firm's product or service. Type 1 companies rely on their personal ties to the university for continuing information flows and library access for literature reviews and patent searchers.

University faculty and researchers are the main source of company personnel. Type 1 firms hire university graduates and employ faculty in consulting arrangements. The professional, yet personal, networking relationship Type 1 companies have with the university helps them greatly in hiring decisions as prescreening of candidates can be done informally through the network. University faculty may also serve on an advisory team to guide the company's technical direction. Often, professors hold stock in the company.

Type 1 firms may use their connection with the university to establish initial business credibility. They may join a university consortium or centre in which the university has a leading role. As a new start-up at the time they join, membership in the consortium can

TABLE 7. A suggested typology of firms

What works?	What doesn't work?
Type 1 firms (small spin-offs	
Licensing	
Close personal ties	University limitations on Professor's consulting time
Consortiums and centres	Large up-front licensing fees
Federal and state small business assistance program	
University willingness to take equity shares	
Type 2 firms (small to medium, stable but growth-oriented)	
Delivery of tangible products or processes	Negotiation of intellectual property rights
Centre affiliations	Lack of secrecy in universities
Federal and state grants	
Type 3 firms (medium to large with strong internal R&D)	
New technology champions	Personnel shifts
Federal grants (ATP & TRP)	
Personal contacts and interactions	Lack of their own internal resources to fully take advantage of what the university has to offer
Planting seeds for future growth	
Type 4 firms (medium to large undergoing R&D down-sizing)	
Technology champions	University culture
Delivery of tangible product or process	
Selectivity	

provide both visibility and credibility to the young company. Their membership fees are likely to be adjusted on a sliding scale so that their initial entry fee is affordable.

The University's fiduciary missions along with state assistance programmes, especially those emphasising job creation, are quite helpful to Type 1 firms. They may have taken advantage of the university's business incubator or small business assistance programmes that exist on campus (especially assistance in writing a business plan).Type 1 companies also work with university scientists to receive government grants (SBIR and ATP).

Since their relationship with university personnel is often one of dependency, Type 1 companies are likely to see university limitations on the amount of time a professor may devote to private consulting as a problem. Since they are heavily concerned with knowledge transfer, on-going access to the informal professor network is crucial to good relationships.

Type 1 companies are normally short on capital. A major obstacle reported by Type 1 firms is insufficient capital for the large up-front payments that universities often ask when licensing a technology. Willingness on the part of the university to take an equity position in the firm, instead of insisting on a high initial licence payment fee, is often considered crucial to Type 1 firms. Hesitance to take an equity share might be considered an anti-small business bias on the part of the university. While the university wants a large, up-front transfer payment for a technology, small firms cannot always comply. The university's lack of awareness of scarce financial resources in small firms (especially new start-ups) and the university's hesitancy to adopt alternate mechanisms of transfer (such as taking an equity holding in the new firm), are barriers to collaboration.

Type 2 companies are small to medium, well-established, less high-tech companies, looking to the university for growth. Interactions with the university may include centre affiliation, licensing a technology, employing university researchers as consultants, providing student internships, library use, technical assistance, and local chapter professional society interaction. Personal interaction is, as with Type 1 firms, a mainstay of interaction. Type 2 companies heavily interact with universities through the vehicle of state grants, which require university–firm partners, and federal grants, such as SBIR, with similar university–firm collaboration requirements.

Type 2 firms tend to define successful interaction with universities in terms of tangible products or processes. The important question for Type 2 firms is: 'Did you bring a product to the market?' Interactions that do not result in a new product or process which will help the company grow are seen as unsuccessful.

Type 2 firms report two primary barriers that arise when they attempt to work with universities. Negotiation of intellectual property rights is reported as difficult. Because Type 2 companies are focused on tangible products or processes as the basis of workable university interactions, they are prone to experience ongoing difficulty negotiating intellectual property rights with the university. They may ask in advance for an exclusive license to any commercial development coming out of joint research efforts—a demand universities may be unwilling to meet. Type 2 companies suggest that the university's too often held view, that all companies are wealthy, is a barrier to interactions. The university argues that the technology they wish to license is close to a commercialisable product while the firm often sees it as very near the idea stage. The result is that the university places a higher value on the technology than the company. This, combined with the belief that the corporation is wealthy and can pay, creates a barrier.

This focus on tangible product or process outcomes of university interaction may also result in secrecy issues as being seen as problematic. Type 2 companies may see the dialogue of researchers in university centres as lacking a full appreciation of firm propriety needs. Researchers in the university's facilities ordinarily share information with colleagues, but Type 2 companies are likely to see this as a conflict. Since these colleagues may consult for local competitors, information flows (without cost) to the competition. From the Type 2 firm's point of view, they have a free rider problem. From the researcher's point of view,,it is merely open communication among researchers.

Type 3 firms are medium to large high-tech firms with a strong internal R&D component. These firms have a strategy of using the university to add R&D capability to the firm. Type 3 firms engage in myriad links with universities. Type 3 companies may have a policy of an internal identification of a new technology 'champion'. This champion actively searches the scholarly literature, attends professional conferences and trade groups, and participates in technology transfer conferences seeking information about this new technology. When this internal champion links with a champion within the university, knowledge and technology flows are facilitated. Type 3 firms generally have their own internal technology transfer liaison personnel who work full-time on establishing and monitoring university R&D relationships. Type 3 companies sponsor university research, license technologies, affiliate with centres, employ university researchers as consultants, and endow chairs. They take advantage of federal programmes, such as ATP and TRP, which have a university–firm collabcration requirement. Type 3 companies attempt to create goodwill with the university by donating money with no strings attached and by entering into free material or equipment use agreements with university researchers.

Many of a Type 3 company's university interactions are driven by informal individual contacts. For instance, company personnel may sit on alumni advisory boards at the university and interact with university professors frequently at professional society functions. The on-staff PhD scientists may teach occasional courses at the university and Type 3 firms may hire university professors to come to their site and offer professional short courses.Type 3 firms hire many of the university's graduates and look to the university for an on-going source of well-trained employees. Since Type 3 companies establish and maintain relationships based heavily on personal interactions, personnel shifts are considered quite problematic. Once the key people move, reestablishing relationships is very difficult.

Type 3 companies ordinarily maintain associations with several universities and use those linkages as part of their corporate strategy. They believe that university interactions give their products exposure (especially when a professor will mention a product in a paper or presentation) as well as giving their internal personnel access to state-of-the-art knowledge flow (and thus an advantage on the next generation of products).

Generally, Type 3 firms use some aspect of the length of the relationship as part of their definition of successful university–industry partnering. While delivering the sought-after technology is at the root of the relationship, Type 3 companies are willing to abide some level of specific project failure in exchange for the overall value of long-term university interaction. Success is not just seen as the production of some tangible product or intellectual property; rather, it is seen as the potential long-term results of the interaction itself. Expectations of university interaction are guided by the concept of throwing out a bunch of seeds and seeing what takes root. The research relationships that do not produce a specified target are seen as public relations investments as opposed to failures. If a good relationship is created, the Type 3 company is satisfied. A Type 3 company's long-term strategy leads them to believe that by sowing the seeds of mutual trust with university researchers today, even if specific current projects fail, they might be the company that benefits the next time the researcher has a winning idea.

In line with this viewpoint, the primary barrier to university–firm interaction reported by Type 3 companies is the internal corporate obstacle of having insufficient capability to take full advantage of what the university offers.

Type 4 firms are medium to large high-tech companies with a substantial internal R&D component under the stresses of downsizing or restructuring. Links to universities are seen strategically as opportunities to fill gaps. Type 4 firms, like Type 3 companies, have internal R&D personnel who scan the professional literature seeking researchers working in areas of interest to the firm. Type 4 firms often maintain many associations with several universities. These relationships include (but are not limited to) serving on university advisory boards, hiring university graduates, centre affiliation, contracting university research, and granting R&D dollars to universities. Type 4 firms are selectively reducing the number of universities with which they maintain contacts. Universities may heavily solicit Type 4 firms for R&D arrangements but Type 4 firms are becoming increasingly selective regarding the universities with which they associate.

Type 4 firms are often not willing to pay university overhead rates and will attempt to bring researchers of interest aboard on a private consulting basis. Type 4 firms will selectively engage in sponsored research and pay standard overhead rates, but generally only if the intellectual property rights are negotiated in advance in terms favourable to the company. Type 4 firms primarily define successful university–industry interaction in terms of a tangible product or process. Type 4 firms suggest that university–industry linkages can also be successful even if no tangible product or process results. If the company is able to direct the research agenda of the discipline so that a problem is tackled (even if it is not solved), then that interaction is successful. Finally, if university–firm joint research projects result in corporate decision-making, then the interaction is considered successful. This definition would primarily apply to failed research efforts. If the sought-after research result is not obtained, internal firm decision-making will be facilitated—that is, they will know they have to go in a different technological direction. Joint university–industry projects can fail but at the same time that interaction can be quite successful if it promotes internal decision-making.

Type 4 firms point to the culture conflicts between the university and firms. Professors are at times seen as not understanding business nor the concept of a customer. The failure of the professors to understand that there must be a relationship between the dollars given to research and product development is a large barrier to university–firm linkages. They tend to see the university system as disorganised when it comes to university–industry linkages and cite this disorganisation as a barrier to interaction.

Type 4 firms also see barriers stemming from the desires of the collaborating researchers to exploit their work for personal gain so that they spend too much time on producing publications and not enough time on the project itself. Finally, Type 4 firms indicate that sorting out the details of intellectual property rights can be a barrier to collaboration.

6. Conclusions

This chapter has used the framework of interorganisational complexity to analyse university–industry technology transfer-oriented interactions. From the university viewpoint, three questions were posed. What characteristics differentiate researchers who engage in technology transfer from those who do not? What are the most important factors that account for the decision of an individual researcher to undertake technology transfer as an activity? What factors explain the extent of technology transfer activity undertaken at universities? From the viewpoint of the business community, a fourfold typology of firms was suggested. The question of what works and what doesn't work was explored for each of the four firm types.

This study suggests that there are real differences in the profiles of research us who engage in technology transfer-related activities and those who do not. Institutions seeking to enlarge the numbers of participating researchers might wish to look carefully at this profile and to modify policies to encourage participation. While providing the institutional structures for collaborative efforts seems effective in expanding the level of technology transfer interaction, the factors influencing the individual decision to undertake technology transfer related activities suggests a different set of incentives that ought to be the focus of policy modification efforts. This study also suggests that universities need to be aware of the differences in company outcome expectations if they are to be more successful in working with firms. Further exploration of the primitive typology of firm types developed here, along with a more complete elaboration of outcome expectations and forms of interaction linked to each firm type, seems a fruitful place for additional research.

Notes

1 The author is grateful to the National Science Foundation for funding this research under grant no. SBR-9305591.
2 The dataset contains questionnaire responses from 254 university-bound researchers, and 759 industry-linked researchers. More than 80 per cent of both university-bound and industry-linked researchers are tenured associate or full professors.
3 Non-parametric correlational analysis reveals a statistically significant and strong pattern. By assigning a value of one to industry-linked researchers and a zero to university-bound researchers, correlations were run. There are strong positive correlations between industry-linked researchers and the researcher taking the initiative to personally contact a firm (tau-b of .51) or being personally contacted by a firm (tau-b of .50). Undertaking consulting activities and contacts from former students both correlate highly with the designation of industry-linked researcher tau-b of .43 and .30, respectively).
4 The correlation between each of the variables (night classes, student internships, professional workshops, satellite courses, classes held at firm locations, and weekend classes) and industry-linked researchers are all statistically significant but weak. The highest correlation (a Kendall's tau-b of .19) is between industry-linked researchers and departments offering professional workshops.
5 Correlation analysis reveals a significant but weak relationship between variables. The two highest correlations are between industry-linked researchers and the existence of cooperative R&D arrangements with firms, as well as the existence of personnel sharing agreements with firms (Kendall's tau-b of .20 and .14, respectively).
6 There is a moderate to strong correlation (tau-b of .33) between industry-linked researchers and having filed for or been granted a patent. Nearly 60 per cent of industry-linked researchers indicate they hold or have applied for a patent as compared to 17 per cent of university-bound researchers. The results are highly statistically significant with a Chi Square test statistic of 137.5.
7 Industry-linked researchers are slightly less apt than university-bound researchers to

believe the university rewards researchers for undertaking patenting activity. The tau-b correlation (of -0.11) is statistically significant but weak. Industry-linked researchers are more likely than others to feel pressured to become involved with applied industrial research efforts because they sense that granting agencies, university, department, or centre administration will look favourably upon such activity. Finally, industry-linked researchers are less prone to consider the recent emphasis on industrial outreach to improve national competitiveness as negatively impacting the university's basic research mission (tau-b of -0.13).

References

[1] Barker, R. (1985), 'Bringing science into industry from universities', *Research Management*, November–December.

[2] Bonaccorsi, A. and Piccaluga, A. (1992), 'A theoretical framework for the evaluation of university–industry relationships', *R&D Management*, 24, pp. 229–46.

[3] Crow, M. and Emmert, M. (1984), 'Interorganizational management of r&d: university–industry relations and innovation' in Bozeman, B., Crow, M. and Link, A. (eds), *Strategic Management of Industrial R&D*, D.C. Heath and Company, Lexingon, MA.

[4] Fairweather, J. S. (1990), 'The university's role in economic development: lessons for academic leaders', *SRA Journal*, Winter, p. 11.

[5] Fassin, Y. (1991), 'Academic ethos versus business ethics', *International Journal of Technology Management*, 6, pp. 533–46.

[6] Frye, A. L. (1985), *From Source to Use: Bringing University Technology to the Marketplace*, American Management Association, New York.

[7] Geisler, E. and Rubenstein, A. H. (1989), 'University–industry relations: a review of major issues', in Link, A. and Tassey, G. (eds.), *Cooperative Research and Development: The Industry-University-Government Relationship*, Kluwer Academic Publishers, Boston, Dordrecht, London, pp. 43–62.

[8] Killoren, R. (1989), 'Institutional conflict of interest', *Research Management Review*, 3, pp. 1–11.

[9] Owen, J. V. and Entorf, J. F. (1989), 'Where factory meets faculty', *Manufacturing Engineering*, 102, pp. 48–71.

[10] Ruscio, K. (1984), 'University–industry cooperation as a problem in inter-organizational relations', in Bozeman, B., Crow, M. and Link, A. (eds), *Strategic Management of Industrial R&D*, D.C. Heath and Company, Lexington, MA.

[11] Smilor, R. W. and Gibson, D. V. (1991), 'Technology transfer in multi-organizational environments: the case of r&d consortia', *IEEE Transactions on Engineering Management*, 38, pp. 3–13.

[12] Stewart, G. H. and Gibson, D. V. (1990), 'University and industry linkages: the Austin, Texas study', in Williams, F. and Gibson, D. V. (eds.), *Technology Transfer: A Communication Perspective*, Sage Publications, Newbury Park, London, New Delhi, pp. 109–29.

PATENTS AND TECHNOLOGY TRANSFER IN PUBLIC SECTOR RESEARCH: *The Tension Between Policy and Practice*[1]

ANDREW WEBSTER and KATHRYN PACKER
Science and Technology Studies Unit (SATSU)
Anglia University, East Road, Cambridge, CB1 1PT

1. Introduction

Scientific research generates knowledge which is expressed in different forms of intellectual capital—papers, qualifications, professional status and so on. These in turn can be translated into other forms, such as grants, prizes, gate-keeper positions and a wider social status beyond immediate professional networks. Scientists working within the public sector research base (PSRB)—such as universities, government laboratories, research institutes and centres—are now under pressure to translate their intellectual capital into more material forms which have tangible commercial value—royalties, licensing income, product sales and so on—for their employer institutions. The pressure to exploit the PSRB reflects an 'innovation anxiety' apparent in all industrialised countries [39], and whereas in the past the exploitation of PSRB inventiveness emerged via many indirect routes, today staying up with the innovation game requires a more targeted research policy, and one whose impact can be felt upstream in the basic research lab itself [3, 5, 10, 13, 20].

In the language of innovation models, the translation of intellectual into commercial capital is seen as a process of technology transfer from the PSRB into industry. A dominant assumption since the early 1980s has been that technology transfer can be improved by increasing the incentive within PSRB institutions to commercialise research themselves, notably by devolving to them the rights to intellectual property (IPR) that emerge from funded research. The rate, efficiency and overall momentum of technology transfer would thereby be increased to the economic benefit of both the PSRB—through royalty or licensing income for example—and the nation as a whole.

Thus, many would argue that patenting activity is a good indicator of effective technology transfer [34]. Indeed, one way of seeing the patent is as a device that translates R (research) into D (development) precisely because legal requirements mean that a patent must specify the utility of the claimed invention and provide an 'embodiment' of the claim itself. In this way a patent is, at least in principle, a device for facilitating technology transfer. It is not surprising then that a policy which devolves to the PSRB the right to hold patents should be one which is assumed to promote technology transfer.

There has, however, been little debate about the relation between increasing patent activity in the PSRB and overall enhancement of technology transfer rates. Typically, it is

J. Kirkland (ed.), Barriers to International Technology Transfer, 43–64.

assumed by policymakers that a growth in PSRB patenting activity must be a good measure of overall technology transfer capacity and delivery. This is understandable given the large number of surveys which report close correlation between levels of patenting and economic strength at both company and national levels [2, 22, 31]. Yet such gross correlations may not translate to the specific circumstances that hold within the PSRB and the role it plays within the wider innovation system. Rather than simply asking how can the PSRB engage in patenting more effectively—the sort of question underlying most government policy on this issue—we need to ask what the actual relationship is between university patenting and technology transfer and whether the IP system is itself equipped to engage the PSRB. By looking at the relationship between the two in this way we avoid the simplistic linear assumption that the technology transfer gap can be filled by moving IPR up the innovation stream to the universities themselves. Instead, we start from the position that the relationship between the IP system and the PSRB is culturally and institutionally problematic, one not simply resolved by developing better legal competencies in universities. We believe that the questions this opens up put us in a stronger position to know whether the result of increasing the PSRB's engagement with the IP system is likely to achieve the desired policy objective of improving technology transfer.

This chapter explores these issues through the first detailed quantitative and qualitative survey of patenting activity within PSRB (primarily the university sector) in the UK, as well as analysis of secondary data sources from elsewhere. We argue here that the increased patenting activity apparent in UK universities and government research laboratories does not necessarily indicate an enhancement of the technology transfer process and that without understanding why this is so we may see in the future a growing tension between those policies which in broad terms encourage linkage between academia and industry and those which encourage patenting within the PSRB. Our argument is not that often voiced by some industrial and academic representatives that the PSRB should stay out of patenting, but that in engaging with the patenting system PSRB institutions confront problems which are not addressed by current policy-prescriptions on IPR.

2. Background: the Growth of Patenting in the PSRB

The last fifteen years has been a period of considerable change in the management of the IPR assets of the PSRB. These have taken the form of changes in the structure and ownership of technology transfer organisations coupled with a much greater pressure—rhetorical, bureaucratic and financial—on the public sector to commercialise the outputs of its research. These changes have, to some extent, been inspired by developments in the US and elsewhere. Thus, in the US there has been a gradual increase in the level of industrial funding of the PSRB such that almost 20 per cent of university research programmes is tied to industrial funding. This has been accompanied by a growth in patenting activity among US universities and, for elite institutions, this has brought a steady increase in licensing and royalty income: the top two—the Universities of California and Columbia —each earned around $30 million in 1993 from patent income, almost a quarter of total university royalties. These trends have meant a parallel development in the role and scale of technology transfer offices and expertise within universities (typically members of the Association of University Technology Managers), sometimes at considerable expense: the 1993 budget for the Stanford office, for example, was $1.5 million.

Before 1980 any IPR that resulted from research supported by Federal grants remained in government hands. The Stevenson–Wydler, and Patent and Trademark (Bayh–Dole) Acts of 1980 gave universities (and other non-profit organisations and small businesses) the ownership rights to patents resulting from government-funded research. In 1984, the US Congress extended these rights to all Federal laboratories and centres falling within the jurisdiction of universities. The US government retains the right to use any patented inventions without having to pay a licence or royalty fee. These changes in the law were clearly intended to promote a faster rate of technology transfer from the PSRB, and within a few years the major research universities had developed IPR policies, guidelines to regulate academic–industry linkage, and offices to promote and administer the transfer of new technologies [13, 21].

Patents filed by universities have increased steadily, facilitated in part by these legislative changes, with the rate of patent filing and the number of university patents as a proportion of total patents doubling between 1975 and 1985 [23], and quadrupling between 1971 and 1989 (from just under 300 to a total of 1200) [9]. Major institutions, such as the Massachusetts Institute of Technology, now file between 60 and 100 patents a year. This growth in patenting activity reflects other aspects unique to the US IP system which may help promote a higher propensity to patent among US academics and Federal Laboratory scientists. Two aspects are worth mentioning: first, the US system operates a 'first-to-invent' rather than 'first-to-file' policy on patent claims which reduces the pressure to file quickly while enabling researchers to build stronger (broader) patent claims. It is also said to enable 'small' inventors to patent their work while first-to-file favours large corporations with the funds to meet rapid and extensive filing costs. (At the same time, first-to-invent can produce subsequent dispute between parties over which was first to invent—so-called 'interference proceedings', and these can be costly). Secondly, the US system allows a 'grace period' of one year from the date of any publications relating to a claim and its actual filing: this enables academics to publish material without losing subsequent patent rights. Costs of filing are unlikely to be a significant factor in determining propensity to patent: costs are broadly similar for first filing in the US and UK, and patentees from both countries try to keep European-wide filing costs down by using the PCT route.

Perhaps one other indicator of the greater momentum behind the commercialisation of PSRB in the US has been the felt need to establish guidelines on possible 'conflicts of interest' that arise when publicly-funded research produces private gains for researchers. The NIH, NSF and FDA have recently prepared guidelines which will give institutions the authority to secure information on the financial holdings and potential gains of individual investigators prior to the submission of grant proposals [1], while the NIH has sought to tighten its guidelines on the distribution of IPR created through private sponsorship of public sector research. There has been no parallel debate in Europe: for example, in the latest UK government document on IPR in academia [25], the conflict of interest issue is never discussed.

In Europe patenting activity has been encouraged at national and international levels, the latter through initiatives such as the European Commission's VALUE programme established in 1991. Its aim is to assist in the evaluation of the patentability of research, to help in the submission of patent claims, and thereby to promote the 'valorisation' of R&D.

Whatever the outcome of these efforts to harmonise the European patent regime, there are considerable differences between EU member states in terms of those PSRB organisations principally involved in patenting [19]. The different patterns relate directly to the distinct 'institutional architectures' of different EU countries, that is the different organisational relationships, competencies and priorities that constitute local science and technology systems. In France, for example, it is the national research laboratories of CNRS (which receive more than 75 per cent of all public funds for research), while university patenting is negligible [16]. This reflects in part the very limited contribution of French Higher Education Institutions (HEIs) to research. In the UK, on the contrary, the number of university held patents is higher in both absolute and relative terms (as per cent of total patents) than any other EU country. Whether specific 'architectures' are more or less conducive to both patenting and technology transfer is a question that still needs to be answered. At this point we focus on the development of IPR policy within the UK as the context within which our survey on PSRB patenting should be understood.

3. The Development of an IPR Policy in the UK

> 'What is happening is that the contractual requirements for carrying out research with industry and commerce is becoming more and more important....If you go back five years contract terms and conditions very often didn't talk about IP.' (UK-based ILO)

In the UK in 1980 there was growing dissatisfaction with the ability of the National Research and Development Corporation (NRDC) to identify and successfully commercialise inventions produced in universities from government funded research work. This was despite the fact that the body had the right of first refusal to any IPR from PSRB. These doubts became focused around the emerging technology of monoclonal antibodies (MAbs) and recombinant DNA (rDNA) reviewed in the Spinks report in 1980 [36]. In particular the failure to patent the MRC funded work of Milstein and Kohl on MAbs at the Laboratory of Molecular Biology in Cambridge was seen as extremely deleterious.

In 1981 NRDC was merged with the National Enterprise Board to form the British Technology Group (BTG), which was charged with developing and commercialising technology from the public and private sectors but only two years later the Thatcher administration announced plans to remove BTG's first option rights on publicly funded research, as part of the government's broader deregulatory, free-market strategy. In 1985 universities were given the right and the responsibility to ensure that publicly funded work was transferred to the private sector, specifically through the protection of IPR. This was not a task most universities had any competence in dealing with and was a change in their perceived role. In order to ensure that they were willing and capable of obtaining patent protection on inventions an interdepartmental committee was set up to review universities' IP activity. This committee, the Exploitation Scrutiny Group (ESG) was made up of representatives from the research councils, the higher education funding councils, from the representative bodies of HEI principals and from four government departments (DES, DTI, HM Treasury and the Cabinet Office). Universities were asked to submit written evidence to the committee to show they had mechanisms in place that would ensure that intellectual property would be protected and commercialised. The committee had fairly meagre resources and could not use more proactive methods in scrutinising HEI's work in this area. Of the 117 institutions invited to respond to the committee, 7 did not provide any sort of response. A substantial proportion (22) opted to maintain an exclusive agreement with BTG.

The ESG continued reviewing HEIs but, by 1991, 23 institutions still had to be scrutinised and another 16 had been given two more years to come up to scratch. In its report of 1991 the ESG recommended its own dissolution but thought the elements of its work —validation, counselling, and monitoring of HEIs and strategic policy nationally—could be taken up by different bodies: validation, for example, might be best left to a representative body of the Research Councils.

In 1992 the issue of the exploitation of IPR in HEI was still seen to be important and potentially problematic. The newly formed Office of Science and Technology published a report intended to disseminate best practice throughout public sector research institutions, including several government institutions which had recently become agencies (giving them a greater responsibility for their own financing and management) [25]. The advice is detailed and sophisticated but, like earlier reports [6, 18], it does little to locate the role of IPR in innovation and technology transfer more broadly.

The OST report on IP was followed eight months later by a much more influential and wide ranging document, the first UK government White Paper in twenty years, *Realising Our Potential* [27]. This included a reorganisation of the Research Councils and a greater emphasis in their new mission statements on the users of research outputs. For example, the mission of the new Biotechnology and Biological Sciences is:

> 'To promote and support high-quality basic, strategic and applied research, and related post-graduate training in biological systems with the aim of enhancing the management of biological resources and their utilisation and interactions with the environment, placing special emphasis on meeting the needs of users of its research and training outputs, thereby enhancing the UK's industrial competitiveness and quality of life' [27; 29].

Although the term 'users' can be defined quite broadly, in this context it is generally taken to mean UK industry and government. *Realising Our Potential* discusses and lays out future plans for schemes to promote technology transfer, such as LINK and the Teaching Company Scheme but again says little about the role of IPR in this context. Patenting therefore is still positioned within this debate as a separate, primarily technical task, which skilled operators undertake but which is to a large extent disengaged from the debate about linkages.

The lengthy policy debate on 'best practice' for IPR was never informed by any published information on how universities were living up to the changing expectations made of them. Basic questions such as how many patents do UK universities file each year are in fact very difficult to answer using patent application information sources (such as 'Impadoc' or 'Derwent'), since university research may lead to patents filed by the university, a university or spin-off company, a sponsoring company or agency or even individual scientists.

4. Survey of Patenting Activity in UK Universities

In order to obtain this information we undertook a national survey of the patenting activity and management of all UK universities [29]. A postal survey covering all universities (including those newly designated as such from the old polytechnic sector) was completed in March 1993, with follow-up semi-structured interviews with research scientists and ILOs, at ten universities plus patent agents, patent examiners (at the Patent Office) and industrialists. In total 50 interviews were conducted.

One somewhat surprising initial outcome of the survey was the difficulty many respondents themselves also had in answering the questions due to lack of readily available information on the number of patents the university owned,[2] or total expenditure on this area. This reflects the problem of managing intellectual property in such open and complicated institutions as universities. An alternative explanation is that even though intellectual property has been on the agenda of the higher education sector for some time, a sizeable proportion of institutions still have not formalised procedures and support mechanisms for dealing with this issue. This explanation is supported by the fact that 20 per cent of the universities in the survey did not have a university policy on dealing with IPR at the time of the survey.

The highest number of patents held by any university was 60, with the total number of patents held by all respondents being 510. The distribution of patents across universities is shown in Figure 1. Of these patents 30 per cent were in the areas which are of special interest to our research, that is medical and agricultural biotechnology and more traditional pharmacology. This number (157 patents) compares favourably with small biotechnology companies, which may have only a handful of patents, but represents only

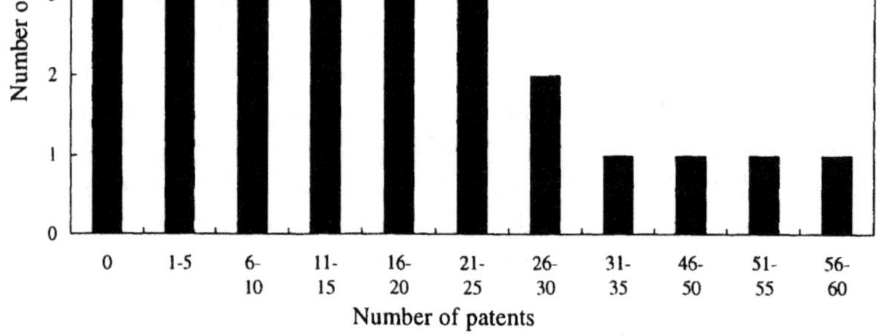

Figure 1 Patents held by UK universities

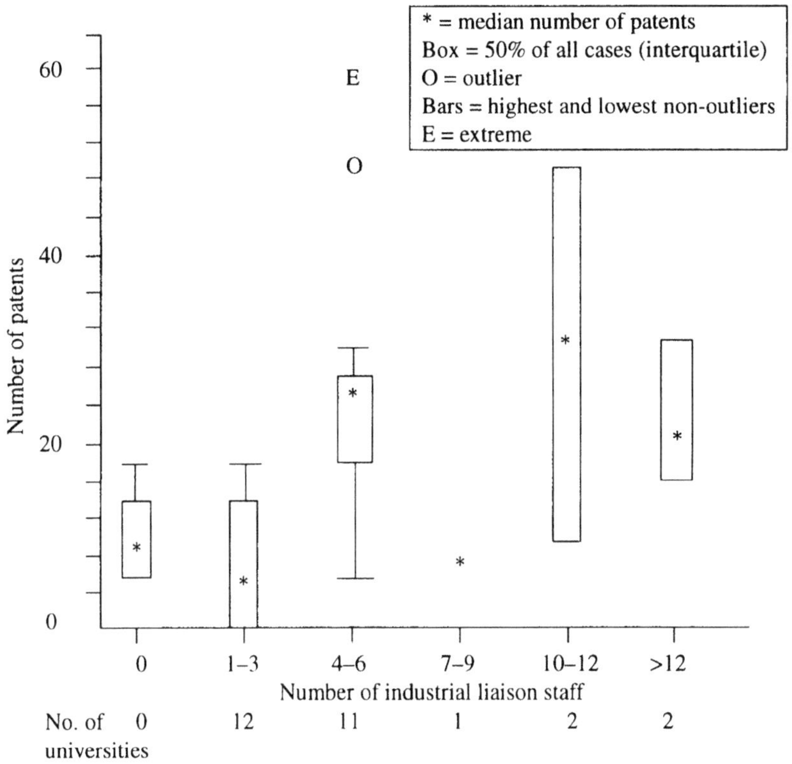

Figure 2. Relationship between the number of ILO staff and the number of patents held

a drop in the ocean of the 10,000 or so patents granted by the UK patent office alone in a single year.

Only twelve universities said that their work on protecting IPR was self financing, although twenty said that it was expected to be so in the future. The mean number of staff of the industrial liaison office, or equivalent, is four (which included a patent agent in only one institution) and the mean age of these offices is seven years. Over 80 per cent of these universities had links with BTG, even after the latter's privatisation in 1992. The relationships between the number of ILO staff, the number of patents and expenditure are shown in Figures 2 and 3. The results do not indicate a strong link between the number of ILO staff and the number of patents, suggesting that although offices may be understaffed this is not the limiting factor where securing patents is concerned. ILOs were generally quite positive about factors such as the increasing awareness of patenting amongst research staff and the faster identification of patentable results but most did not think that this had been translated into more patents per project or more targeted applications. When it came to defending patents in legal disputes resulting from infringement or a challenge to the patent only two universities said that they had been involved in a dispute of any kind.

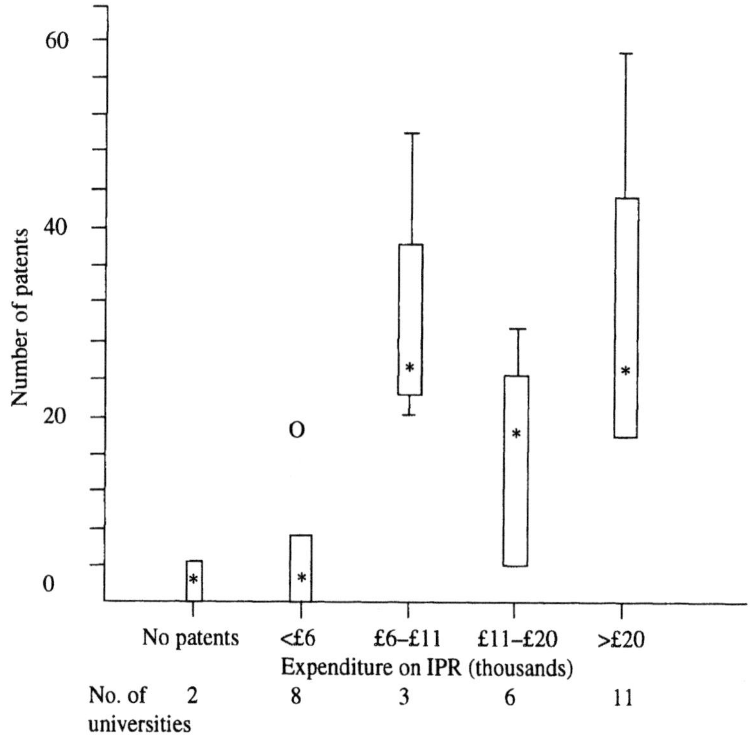

Figure 3. The relationship between expenditure on IPR and the number of patents held

The contribution of intellectual property to university funds was also quite modest. Only two universities had more than ten patents, in the areas of pharmaceuticals and biotechnology, licensed out to companies, although 50 per cent had up to ten licences. Four universities were receiving a significant amount of money, over £50,000, from these licences, whilst six got less than £1,000, five with up to £10,000 and another four with between £10,000 and £50,000.

In summary one can say that, although some universities are doing very well out of patenting, there is considerable variation across institutions and the total number of patents actually held by universities is still relatively low. Many universities are still developing their practices and procedures in the area of patenting. Along with the considerable time lag between patenting initiatives and the receipt of royalties, this should caution us against being too critical of universities which do not appear to have yet 'come up to speed' in the area of patenting. Moreover, our research suggests that 'modest changes in current practice' are unlikely to lead to a major improvement in performance: this understates the cultural and organisational resources and competencies that are needed to translate scientific findings into patentable claims [30]. There are in addition a number of demands made on universities related to devolved IPR which impact on effective technology transfer, and which are discussed below.

5. Devolved IPR and Technology Transfer

'Twenty years ago they just used to say, if they had an invention, "Oh I've got a brother-in-law in Glaxo", and away its gone!'

'As far as the general academic departments are concerned, I am fairly confident there that all IP matters are directed through me.'

These two comments from two industrial liaison officers at different UK universities reflect something of the cultural shift that the devolution of IPR to universities has created. While the first suggests a rather idiosyncratic technology transfer policy based on kin-ties, the second tells a rather different story of university property being monitored and if necessary protected by the centre. The new policies of devolved IPR are, then, premised on the belief that greater control over IPR at the centre coupled with the potential financial returns on patenting will enhance technology transfer.

Whether this is true depends on the capacity of university staff to cope with the demands that patenting activity places on them, demands which relate to;

- occupational role expectations
- the relationship between patenting and research programmes
- the financial costs of patenting
- the impact of patenting on external links
- the impact of patenting on the disclosure and dissemination of research results.

Let us consider each of these in turn.

5.1 OCCUPATIONAL ROLE EXPECTATIONS

Patenting research results is still regarded as an unusual activity for many university scientists, and both they and ILOs have to determine who has primary responsibility for this task. Many researchers commented on the time and effort needed to file and pursue patent claims, a new responsibility which competed with the growing pressure to produce published material for research quality ratings by national funding agencies: as one said,

'It's quite difficult running a lab and getting publications out, which is what you need to do to survive and you need a lot of energy to divert off and file a patent.'

Given the need to 'divert' energy in this way, some ILOs noted that there could be a problem in encouraging researchers to take an interest—'most people will say that it's not what they signed on for'. Perhaps not surprisingly, those who have had a track record of linkage with industry seem more prepared to spend time and energy on taking new ideas through to patent: as one ILO remarked,

'Seventy-five percent of people who tend to come forward with ideas are those who already have reasonable links with industry, through contracts or previous work experience.'

The additional pressures on time and energy that patenting demands means that researchers are keen to get the support of ILOs, and good patent agents: this helps them to demarcate their own area of responsibility and can limit the time spent on patenting. Links with companies can also help to reduce the pressure on scientists to generate the additional time needed for patenting. However, in some areas, such as biotechnology where new technology-based firms are common points of contact, university scientists may find they have to devote more time to assisting the firms take up their inventions. This may even work against the transfer of PSRB inventiveness: thus, as one biotechnologist observed,

> 'The downside of transferring something to a company is then they have to make a product and there is a huge amount of effort goes into the product....It has been a two edged sword transferring a piece of technology to a young developing company, because they needed so much help to work it through that it actually detracted from new ideas.'

The assumption of course is that scientists will be prepared to put in this additional effort precisely because of the financial benefits that will accrue to them via patenting. As one ILO remarked, 'they have every reason to be concerned about the time. If they are not interested in the financial return from the process then there is really no point'. As we saw earlier, however, the financial return on patenting is typically very modest, and indeed this led some scientists to adopt a sceptical view towards patenting: as one said,

> 'Because of the time taken for little reward for the department, patenting of any of the scientific material we are dealing with is not high on my list of priorities.'

Even with successful patents that generate revenue, scientists often believe that they will receive little personal financial benefit from their labours, since the university or external licensee may take most of the proceeds: as one noted,

> 'There is no great stimulus for people like us to go in for it really, basically it's a question of what am I going to get out of it? Just a lot of hard work and then the [charity] and the University will take all the money.'

Against this, a number of universities had recognised that the level of reward allocated to academics for successfully commercialised IP was too low to act as a genuine incentive. In the late 1980s there was a general improvement in the distribution of income that would go to academics, typically on a sliding scale. One university, for example, allows inventors 100 per cent on the first £2,500 through to 33 per cent of any income in excess of £50,000 net. Despite such changes, the general hope and assumption that financial incentives will drive scientists to patent seems, in light of the remarks above and similar observations made at interview, to be questionable. In fact, we find scientists are often prepared to undertake patenting for non-financial reasons. For example, one scientist reported that the primary function of his patent 'was really just to block anyone else using this system'. We shall return to the issue of defensive patenting in PSRB later.

5.2 THE RELATIONSHIP BETWEEN PATENTING AND RESEARCH PROGRAMMES

The management of IPR also poses new questions for ILO and research staff in connection with the overall focus and management of their research. A key question, for example, is how far a department or university should try to develop a convergent set of patents—a strong patent portfolio—rather than a divergent range of patents covering a more varied but perhaps lower commercial potential. A convergent portfolio may enable inventors to secure a dominant set of patents which might provide them with a stronger position in any subsequent moves to exploitation. However, such a proactive approach to patenting areas might be regarded by scientists as a distortion of their research direction —a 'filling in of technology gaps'—which could create ill-will. Such a strategy is a high-risk venture anyway given the pace of technological change and markets, and in most institutions would require a level of direction from the ILO which would be regarded as improper. It is clear from our survey, however, that an increasing number of ILOs are adopting a more selective approach to the sort of patents they will pursue, not least to keep patent costs down, a point returned to below.

This sort of planning is not totally alien to research scientists. They have to position their work in relation to competitors and collaborators, both in terms of experimental programmes and how the work is presented in conference and journal papers. Whatever one says about the exponential growth of scientific literature, there is still a sense that these outputs are in some way meaningful in terms of the judgement of academic credibility of the group and that other people's outputs can contribute to their work. This cycle of information dissemination and allocation of credit has been seen to be at the heart of the functioning of academic scientific institutions. Reading papers not only gives scientists information about others' work, it locates it within a research strategy and teaches less experienced researchers about how to present their work.

How does patenting fit into this? This is a question that we have considered in much more depth in another paper [30], but to summarise, we have found that two thirds of the scientists we spoke to did not use patent literature as a source of information. Half of these had not considered it and the other half had found the literature difficult to access or unhelpful when they had located relevant patents. This is in spite of the fact that the UK Patent Office claims that material published in patents is not often published elsewhere and that patent literature represents an important and under-utilised resource.[3] If scientists do not find patents useful and are unfamiliar with how they are written, then it is not surprising that identifying patentable material and writing a patent (let alone creating a portfolio of patents) is seen as a difficult task. Our data suggest that this is in part a result of inadequate training of students in the patent system, and in part a reflection of the inaccessibility of patent literature (physically and in terms of the way patents are written as 'texts').

A second broad issue a university must address is that any patent claim will require research staff to engage in 'reduction to practice' experimental work and extrapolating the claim to broaden its scope—such as covering a range of plant varieties for example, or analogues of pharmaceutical compounds. In extreme cases—which we did come across —this would mean devoting staff to undertake a series of often tedious, intensive experiments over a period of months to produce the data on which the claim will stand. Again, this would be a strategic choice which would effect the distribution and focus of intellectual labour deployed in a laboratory. Scientists and research centre managers and ILOs clearly need to decide how far they are prepared to go down this route. In some plant science research, this issue is particularly problematic insofar as research is often based on 'model systems' plant varieties which may not be easily reducible to practice in crop varieties grown for the commercial market.

Thirdly, research programmes that are linked in part to industrial sponsorship or contracts generate both 'foreground' and 'background' IPR. The boundaries between these two need to be clarified as early as possible such that research staff and ILOs are sure which elements of the programme are part of the general research stock of the university (background) and which tied into specific research links with industry (foreground). In some circumstances—notably open-ended research funding by single company sponsors —it has been shown elsewhere that boundaries are blurred, with the result that ownership rights are uncertain [38].

Fourthly, the areas in which universities are most likely to have patents are those associated with science based industries. They are therefore likely to be those areas with which there are most uncertainties with regard to IPR. Biotechnology and computer software are excellent examples of this. Although these areas may well settle down and become easier to deal with than at present, it seems probable that there will be other developments of a similar nature. Patent law takes a very long time to change. Even after a sufficient consensus for change has built up the process of international negotiation may take just as long. So one would expect the law to be constantly at least five to ten years behind the science. There is a parallel mechanism for change, through case law. However, a considerable amount of time passes between a patent being filed and a dispute about its validity being settled. This can affect the actual judgement of the case itself. So universities must not only learn the technicalities of the system but have some mechanism for keeping abreast of the changes within it. They should not see these structures and policies as set in stone, so need themselves to manage and exploit research with the sort of flexibility with which they are not normally associated.

Finally, one of the provisions of patent law is the so-called 'experimental use exemption' which allows researchers to use materials and procedures that others have already patented without paying licence fees so long as that use is purely for experimental work alone. This is designed to facilitate the dissemination of patented invention to a broad range of users and thereby to encourage further innovation. However, although universities are regarded as the principal beneficiaries of this exemption the benefits it brings seem to be under threat. PSRB activity also represents part of a sizeable market for research equipment, techniques and reagents, which has been the first target of those seeking to commercialise biotechnology developments. Patent protection, of what might in the past have been freely available techniques, means that companies can object (in principle and in practice) to scientists making and using their own versions of commercially available products. This has been most apparent in disputes over the pricing of PCR marketed by Perkin Elmer under licence from Hoffman La Roche. The market for PCR as a research tool was $26 million in 1990 [33]. One scientist complained in a letter to *Science* that the costs of buying PCR products from Perkin Elmer, rather than producing them very cheaply themselves, would be six times their annual research budget [35].

However commercialisation of research methods is not the only problem, since the more universities take on a commercial function the more their research might be regarded as less than purely 'experimental' and the less likely they are to be seen as innocent upholders of academic freedom. Patent holders, in other universities, may become much more cautious about allowing exemption to research teams in universities whom they know are likely to be encouraged—especially in light of the UK Research Councils' Mission Statements prioritising wealth creation—to commercialise any new findings that emerge from the work. Considerable tension can emerge between universities over this, as has happened, for example, in the case of patenting of the genetic markers for cystic fibrosis, held by universities [8].

In various ways, then, research programmes are affected by the engagement with the patent system, an engagement which not only imposes new managerial demands on staff but also carries potential intellectual costs in regard to how and where they deploy their research energies.

5.3 THE FINANCIAL COSTS OF PATENTING

In devoting energies to patenting, universities and their staff are also bearing direct and indirect financial costs. As we saw earlier, the contribution of IPR to university funds is typically quite modest. The cost of patenting simply in financial terms is initially (at first filing) relatively modest too. The full costs can be split into four parts, the formal cost of applying for patents (filing, search and examination £300 via the UK PO, the cost of patent agents (about £1,000 on average) and other experts, the potential cost of litigation and the opportunity costs of patenting (see Figure 4).

The renewal and any prosecution costs of patents, especially as numbers of filings increase, can rapidly eat into departmental and/or central funds. In one case, for example, a department was having to carry a £60,000 debit on its patenting activity: as one ILO remarked, 'the quickest way to go bankrupt is through patenting'. Precisely because of this cost, many thought that financial risks are best carried by companies rather than

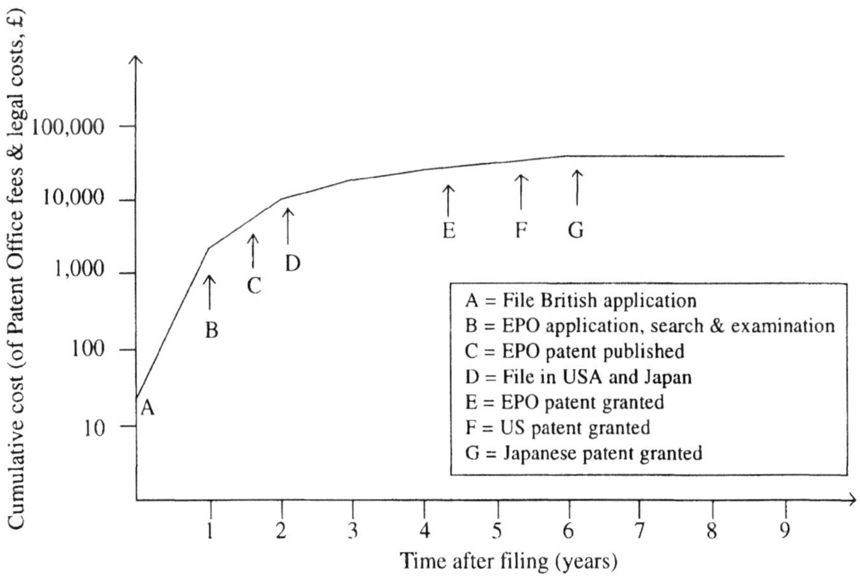

Figure 4. The costs of patenting

the university, especially after the first twelve months. Indeed, ILOs are reluctant to set targets on levels of income that patent portfolios may generate, and certainly most doubt that their offices will be self-financing in the near future. In fact, ILO staff are increasingly having to make sure that their role as income generators (via IPR, short courses and so on) is actively managed in relation to the growing pressure to keep up to a range of 'quality' standards set by a number of external funding agencies, most notably the Higher Education Funding Councils (HEFC). Researchers must ensure that they publish sufficient number of 'quality' papers in academic journals in order to secure a good rating in the four-yearly HEFC Research Assessment Exercise. In real terms, for many scientists the returns on this will be higher than that generated by any patent they file. Therefore, if publications are delayed too long because of patenting constraints and miss inclusion in the cycle, the indirect financial costs of patenting can be high indeed.

5.4 THE IMPACT OF THE DEVOLUTION OF PATENTING ON EXTERNAL LINKS

Prior to devolved IPR, the two more important external agencies for the transfer of technology to the universities were BTG and private firms. We saw earlier that, even with devolved responsibility post-1985, many universities have continued to use BTG's services as a patent and licensing agency which holds over 8,000 patents and returns about £6 million to universities each year in royalties. In addition to this it funds universities to conduct research on its own behalf: in 1990 this amounted to £85 million. In 1992, BTG was privatised (via a management buy-out) and has since sought to secure new patents that will generate revenues in the short to mid-term before potentially high-income patents come on stream in the future. University respondents, not surprisingly, believed that BTG had become more 'hard-nosed' in its evaluation of proposals to patent which it now receives compared with the past when it was suggested the rate of acceptance by BTG was higher. In part this was perhaps because in the past BTG's former monopoly of university IPR (derived from research council funding) had allowed it to build patent portfolios from a range of different (university) sources to produce a commercially valuable set of claims in a specific technological area. After privatisation, BTG was said to be more reluctant to take on material which had too small a potential. According to our respondents, this appears to have helped encourage the drafting of better quality patent claims as well as encouraging the development of stronger links with local technology transfer brokers, or indeed the establishment of university holding companies through which IP is channelled. Accompanying these changes has been the recognition that more attention needs to be devoted to marketing and licensing technology if patenting is to be effective. Most ILOs have tended to shift from a technology push to a market pull approach as a result, and some scientists appear to recognise this: as one said, 'to really succeed in this business you've got to be really close to the market place....otherwise you're not going to make anything that's saleable'.

These developments can however throw up new problems, not only in terms of new organisational demands internal to the university, but also in terms of links with companies outside. While patenting can, as one respondent put it, 'show industry that we [the universities] are serious', it does mean that universities can present themselves as less accessible to firms and even as competitors to them. Clearly, to avoid this, they need to develop clear policies on such matters as when to grant exclusive and when non-exclusive rights and how best to license such rights. If exclusive rights to exploit university IPR are extended to a firm the company might feel it is in a stronger position to take the idea through to the market place because it would not have to share the invention with other (competitor) firms, or engage in cross-licensing with third parties. At the same time, universities need to ensure that their background rights are clearly defined such that they can be, where appropriate, licensed on a non-exclusive basis to keep open a range of possible research and development paths in the future.

More importantly, in some situations the desire of universities to show industry they 'are serious' can work against effective technology transfer. Some companies have pointed out to us that premature filing by universities keen to show the commercial relevance of their work can in fact compromise the future potential of that work. This is because the full value would emerge only with further research work, and early patent publishing allows competitors to engineer around the patent more quickly than if the idea had been kept secret for a longer period. This criticism of premature patenting was also voiced during the recent controversy in the US over the filing of human gene sequences resulting from work on the Human Genome Project [11]. Universities are clearly in a difficult situation here since they believe that many companies will be reluctant to support research if there is no patent to attract them, but too early a patent will reduce the potential IP value and so reduce the chances of securing industrial funding.

The difficulties arise here precisely because the devolution of IP creates the need and demand for more formalised frameworks for working with external organisations in the private and public sectors. While this is a necessary development, it is a moot point whether the growth of formal structures will facilitate any real net growth in interaction between academia and industry and so the potential for technology transfer. One could argue—as suggested with 'premature' patenting—that in some situations, the growth of formal in-house procedures associated with commercialisation might either merely replace what was once done by industry or actually constrain linkage. There is good evidence that informal networks are more important in providing links than formal ones [15, 37]. We shall return to this point later.

5.5 THE IMPACT OF PATENTING ON THE DISCLOSURE AND DISSEMINATION OF RESEARCH RESULTS

Finally, it is apparent from our data that external links which depend on the traditional sharing of information and materials for experimental purposes can be constrained by the growth of patenting in universities. A number of our respondents said that they would now be more careful about disclosing too much information in their papers. For some this is to be expected, particularly those tied into a commercial contract with a company which would vet any publication as a matter of course. But others also suggested that they would be careful about the way they prepared the discussion sections of their academic papers lest their speculations as to the implications of their work could subsequently be deemed to be 'prior art' in the public domain and therefore preempt future patent applications. Indeed one research group found to their surprise that they had been 'caught out' by their own previously published papers. As the senior researcher observed,

> 'We don't write as quickly and we certainly don't put in statements that could come back to haunt us... We've been caught by our own material as prior art. But that changes your attitude.'

And another respondent who had had this problem observed,

> 'I've got a paper on my desk at the moment that I am carefully doctoring the title and several parts of it to make sure it doesn't interfere with what we're about to patent next month.'

Here again we see an emerging tension developing for university staff: as the quality and significance of their research as academic work is increasingly judged on where and what they publish, any restriction on this because of the need to be guarded for patenting purposes will require a carefully synchronised publication of papers and patents to minimise the damage and maximise the return to each. On the other hand, a small number of respondents had also recognised that what is patentable may not be publishable, not because of the latter jeopardising the former, but because some claims acceptable for patenting are of such small significance that they would not be worthy of a paper. One would still have to be careful about public disclosure in such circumstances but at least one would not have to cope with the pressure to disclose via an academic paper. The opportunity costs here are clearly lower.

The need to guard one's intellectual property has, of course, been a commonplace of normal competition among professional scientists. But the engagement with patenting has clearly increased this caution in a different way. It is interesting to note that some scientists use patent filings to guard their own research activity from being constrained by other researchers and, paradoxically, by companies. This is so contrary to the ostensible purposes of patenting as a facilitator of technology transfer that it is worth quoting a particularly clear example of this in full:

> 'Our work is definitely not commercially oriented, it is academically oriented, but every new thing we find is going to be latched onto by the big drug companies, and they might even prevent us from carrying on. If we feel this is a big prospect we will take out patent cover.'

Here then, patenting is undertaken explicitly to ward off what is seen as predatory or unwanted interest from companies. Clearly, therefore, there are some circumstances when devolved IPR can be used to restrict exploitation by external agencies. This was presumably not the intention behind the policy changes during the 1980s. Furthermore, there are aspects of the IPR system which relate to its legal and legislative framework which are currently problematic for anyone (individual or organisation) engaging with it, but which pose special problems for actors in the PSRB.

One of the key issues here relates to the debate around the first-to-file versus first-to-invent issue mentioned earlier in our paper. The legal debate swings around the possibility of international standardisation of patent law around first-to-file rather than first-to-invent. This largely relates to the USA, although the Japanese have a grace period which is slightly different to the position of the European Patent Convention (EPC). Until recently it appeared that the USA was heading towards first-to-file in harmonisation with patent law elsewhere. This was being resisted as it was seen as making things more difficult for university researchers to patent, however it might well have helped some US academic patent holders, who had been missing out on protection via the EPC because of publishing before filing. While some of the ILOs in our survey and interviews did suggest that a first-to-invent system would be beneficial to their universities and might mean an increase in patenting in the PSRB, others disagreed and thought first-to-file avoided much of the litigation or interferences in the USA based around disputed invention dates. A first-to-invent system also means that much greater weight is placed on documentation of research work in lab notebooks or through registration of inventions with central administrators, and the introduction of such a system would require greater attention to these practices than is currently afforded in the UK.

6. Discussion

We began this chapter by reviewing the policy changes that have been made over recent years to promote patenting activity within the PSRB. As we noted, this has been based on the assumption that improved patenting will facilitate better exploitation of the science base and strengthen ties with industry. We saw too how the absolute level of patenting has indeed increased, especially in the life sciences. However, a relationship between patenting in the PSRB and wider industrial competitiveness has never been formally established, and we have shown in this paper that the relationship is likely to be complex and certainly much more problematic than government policies have recognised.

Moreover, the statistical data that show that rates of patenting are high in certain research areas, such as pharmaceutical research, is not an indicator *per se* of higher rates of technology transfer but of the form of control over IPR regarded as crucial to those areas: patenting in pharmaceutical research is much more to do with the nature of control over the specific scientific and technological know-how embedded in the claim than it is to do with rates of technology transfer [4]. Thus PSRB fields which have lower patenting rates—such as engineering—do not thereby have a lower technology transfer capacity.

Our general argument, therefore, leads us to the conclusion that the relation between patenting and technology transfer is complex and in some circumstances less than mutually supportive. Moreover, devolved patenting is supposed to give universities more control over technology transfer through which they can secure additional income from licensing to the private sector. But it might be suggested that more patenting activity by PSRB institutions will simply clog up the technology transfer system with more ideas chasing inadequate long-term financial investment. One industrialist has recently claimed that 'the very small return generated by considerable efforts by highly professional technology transfer specialists [in universities] reflects a general perception by industry that IPRs are not of themselves valuable. It is selling products that makes money, not licensing ideas, and IPRs are at best a means to that end' [12].

The second half of this chapter discussed five main demands which universities must cope with when engaging with the IP system. Meeting these carries various costs relating to occupational expectations, forms and levels of reward from other sources, and additional experimental labour. We also suggested that, ironically, engaging in patenting may—intentionally or unintentionally—constrain the relationships universities have with private sector firms. Relatedly, it would be useful to know whether the formalisation of links between academia and industry that is reflected in higher levels of PSRB patenting constrains the looser, informal ties between researchers in the public and private sectors. Recent work [14] has shown how important informal links are.

The problems this second half of the chapter raise have not been properly addressed by the policy changes made over the past decade or so, yet they clearly determine the context in which universities are seeking to exploit their intellectual property. The policy shifts are also based, we believe, on a particular interpretation of the general relationship between patenting and innovative competitiveness which recent work by economists has challenged: that is, the new policies assume that, just as patenting is central to industrial R&D and competitiveness, so anything that can be done to encourage patenting in universities can only add to the overall competitiveness of the national innovation system. In fact, work by Griliches [17] has shown that patents are a poor measure of innovative activity in firms inasmuch as the amount of unpatented research and development is surprisingly high; moreover, the key issue with regard to the importance of patents is not so much their origin or number but who uses them and the way they are used. Others have shown that firms can be competitive without engaging in patenting [32].

At the same time, encouraging universities to engage in more patenting may be based on a distorted view of how companies go about this process themselves. That is, the policy might be said to understate the effort that patenting can involve for industry. In companies where patenting is significant, notably in pharmaceutical firms, the level of resources—financial, human, and technical—that is given to monitoring and securing intellectual property is much higher than any university could mount. Such firms will also seek to build a portfolio of patents around a particular technology, and to secure this may file up to 300 patents each year, being prepared to abandon over half of these at the end of the first year after filing if they look less than promising.

Apart from exploring the demands placed on universities, we have also considered in the latter part of the chapter whether the IP system is itself equipped to meet the peculiar demands of the PSRB. Thus, we noted, for example, how IP law can often lag behind developments in basic research emanating from the PSRB, how uncertainty surrounds the 'experimental use' exemption and how the debate over first-to-file versus first-to-invent has particular implications for academic scientists.

In response to these difficulties, some might argue that the IP system needs to be modified at the margins to accommodate the particular needs of the PSRB. Others might adopt a more radical view and claim that there is a fundamental difference between the research contexts and requirements of the PSRB and industry, and between their roles within the wider innovation system, roles which are confused and blurred by the introduction of patenting to universities' institutional agendas. Indeed, there have been a number of commentators who argue along these lines. Nelson, for example, has argued that, 'to try to make universities more like industrial labs will tend to take attention away from their most important functions, which are to be a major source of new public technological knowledge and societies' most effective vehicle for making technological knowledge public' [24, p.240]. A key term here, of course, is 'public'; scientific knowledge has typically been regarded as a 'public good', which as codified, generic knowledge is available to all interested in accessing, understanding and using it, and not merely scientists themselves. To alter the terms on which it is available, through patenting, might jeopardise its function as a 'public good'.

This is, however, a position which, at least put as baldly as this, we believe can exaggerate the differences between the two sectors, while failing to acknowledge their similarities. As Callon has argued elsewhere [7], academic science is never solely a public good in the sense of being openly available to 'all': instead it is the 'property' of those researchers that produce it, and that by the very process of disseminating it to other researchers confirms their authorship and receives the recognition this authorship brings. Intellectual property of this type brings reputation, credit and associated symbolic and material (for example, grant) rewards, just as the appropriation of intellectual property via patenting secures rights over a technology to be appropriated via licensing or direct sales of the product. Both tracks also require the disclosure and dissemination of knowledge.

It is not then the non-ownership/ownership or the non-appropriation/appropriation of intellectual property that provides the essential distinction between public and private science. Nor is it likely that we can always distinguish 'public' from 'private' labs in terms of the substantive questions that each addresses: in some fields, in particular, not only can we find increasing communality of research interests as linkage between the two grows [38], we can also find that the acquisition of machinery, instrumentation and the standardisation associated with this equipment ensure laboratories in both sectors experience what economists elsewhere would describe as a form of technology levelling.

Some might argue that it is their character as state-funded activities that makes public sector research distinguishable from private. Even this is inadequate, however, since private sector research and development often relies on direct state investment, such as in the defence industry.[4] Nor can we simply say it is the relative costs for accessing information that distinguishes the two, since it is not so much accessing information but using it that determines real levels of expenditure in either sector.

These arguments suggest to us that it is perhaps impossible to construct a stable and objective distinction between the 'public' and the 'private'. Perhaps it is better to view these as vocabularies of rhetoric or discourses that are mobilised by social actors and organisations: what counts, therefore, as being public—in the 'public interest', a 'public good', the 'public domain' and so on—reflects the capacity of groups to establish that thing, that technology, that policy-issue or whatever, as being a matter which it would be illegitimate to regard as anything other than a matter of collective interest. Such matters are usually associated with forms of accountability and regulation laid down by governmental statute—as in health and safety at work provisions, for example.

However, in the case of the patenting of university science, here the public–private domains have, it appears, been deliberately blurred. Thus, the British government speaks of the contribution of the university science base to 'UK Ltd', or 'UK plc' where the national interests of the United Kingdom are identified with private enterprise. Now scientists contributing to the 'public good' are encouraged, via patenting, to engage in private enterprise through their universities. Yet at the same time these same scientists, through a more and more formalised, external review, assessment, and auditing of their work, are being asked to deliver on 'quality', output of publications and so on. There is, we suggest, a potential conflict here which will continue to cause institutional and professional uncertainties for scientists in the PSRB.

7. Conclusion

This chapter has reported on the first detailed survey of university patenting within the UK and has explored a range of issues that scientists must confront and resolve in endeavouring to patent their work. Our evidence suggests that we should be wary of believing that heightened patenting activity will lead to improved technology transfer between academia and industry. A policy orientation which assumes that difficulties in engaging with the patent system can be resolved simply by improving the training of scientists in intellectual property law and in the filing and prosecution of patents, fails to address the complex processes we have outlined here. Among the various policy initiatives taken to allay the current 'innovation anxiety', the devolving of patenting to universities may merely engender anxiety of a different nature within the science base.

Notes

1 The research reported on here is part of a UK Economic and Social Research Council funded research project (ROOO 23 3302), 'Patenting in academia and industry: determining novelty and scope in two distinct R&D sectors'.
2 University-owned patents are ones where the patent is assigned to the university, or university-owned company, rather than to an individual member of staff or unrelated company or organisation.
3 A leaflet publicising the UK Patent Office Roadshow, which provides an introduction to intellectual property rights, invites people to 'Find out how the use of patent information, 80 per cent of which is not available from any other source, could save up to 30 per cent of your research and development costs' [28].
4 In 1991–2 the proportion of UK government funded R&D going to higher education institutions was only 27.4 per cent, where money going to private enterprise made up 32.2 per cent, with most of the rest being spent by government departments themselves [26].

References

[1] Anderson, C. (1993), 'White House seeks uniform policy', *Science*, 261,17 September, p. 1516.
[2] Archibugi, D. (1992), 'Patenting as an indicator of technological innovation: a review', *Science and Public Policy,* 19, 6, pp. 357–68.
[3] Berman, E. (1990), 'The economic impact of industry-funded university R&D', *Research Policy*,19, pp. 349–55.

[4] Bertin, G.Y. and Wyatt, S. (1988), *Multinationals and Industrial Property*, Wheatsheaf, Hemel Hempstead.
[5] Bozeman, B. and Crow, M. (1991), 'Technology transfer from US government and industrial R&D laboratories', *Technovation*, 11, pp. 231–45.
[6] Cabinet Office (1983), *Intellectual Property and Innovation* (Nicholson Report), *Cmnd* 9117, HMSO, London.
[7] Callon, M. (1994), 'Is science a public good?', *Science Technology and Human Values,* 19, pp. 395–424.
[8] Coghlan, A. (1993), 'Vital research caught in patent crossfire', *New Scientist*, 23 January, p. 4.
[9] Ditzel, R. G. (1991), 'Public law 96–517 and risk capital: the laboratory market connection', *The Journal of Association of University Technology Managers,* 3, pp. 1–23.
[10] Dorf, R. C. and Worthington, K. F. (1990), 'Technology transfer from universities and research laboratories', *Technology Forecasting and Social Change,* 37, pp. 251–66.
[11] Eisenberg, R. (1992), 'Genes, patents, and product development', *Science,* 257, 14 August, pp. 903–8.
[12] Elliot, C. (0000), 'Intellectual property rights—vital tool or unnecessary hindrance?', paper presented at 'Commercial Opportunities for University Research: The Role of Technology Audit', 2-3 December, Heathrow, London, UK.
[13] Etzkowitz, H. (1989), 'Entrepreneurial science in the academy: a case of the transformation of norms', *Social Problems,* 36, pp. 36–50.
[14] Faulkner, W. (1994), 'Conceptualising knowledge used in innovation: a second look at the science–technology distinction and industrial innovation', *Science, Technology and Human Values,* 19, 4, pp. 425–58.
[15] Faulkner, W. (1994), 'Making sense of diversity: public–private sector research linkages in three technologies', *Research Policy,* 23, pp. 673–95.
[16] Geuna, A. (1994), Private correspondence, based on MERIT research.
[17] Griliches, Z. (1990), 'Patent statistics as economic indicator: a survey', *Journal of Economic Literature,* 28, pp. 1661–707.
[18] (1988), *Interdepartmental Intellectual Property Group Intellectual Property Rights in Collaborative R & D Ventures with Higher Education Institutions*, HMSO, London.
[19] Malerba, F. (1994), 'Research interfaces in Europe: the role of innovation systems, institutional architectures and technological competencies', Paper presented at 'University Goals, Institutional Mechanisms and Industrial Transferability of Research', 18–20 March, CEPR, Stanford University.
[20] Mansfield, E. (1991), 'Academic research and industrial innovation', *Research Policy,* 20, pp. 1–12.
[21] Matkin, G. (1990), *Technology Transfer and the University*, Macmillan, New York.
[22] Narin, F. (1994), 'Patent bibliometrics', *Scientometrics,* 30, 1, pp. 147–55.
[23] National Science Board (1987), *Science and Engineering Indicators*, Washington, DC.
[24] Nelson, R. R. (1989), 'What is private and what is public about technology?',

Science, Technology and Human Values, 14, 3, pp. 229–41.

[25] Office of Science and Technology (1992), *Intellectual Property in the Public Sector Research Base*, HMSO, London.

[26] Office of Science and Technology (1993), *Annual Review of Government Funded R & D*, HMSO, London.

[27] Office of Science and Technology (1993), *Realising Our Potential: A Strategy for Science and Technology, Cm* 2250, HMSO, London.

[28] Patent Office (1993), *The Patent Office Roadshow*, Publicity leaflet.

[29] Packer, K. (1994), *Patenting Activity in UK Universities: Results of a National Survey Industry and Higher Education,* 8, 4, pp. 243–47.

[30] Packer, K. and Webster, A. (1994), 'Patenting culture in science: reinventing the scientific wheel of credibility', SATSU, mimeo.

[31] Patel, K. and Pavitt, K. (1991), 'Europe's technological performance' in Sharp, M. and Walker, W. (eds.), *Technology and the Future of Europe*, Pinter, London.

[32] Reid, G., Siler, P. and Smith, J. (1994), 'Quality of Patenting in the UK Scientific Instruments Industry: Database Construction. The Role of IPR in the Innovation Environment', An ESRC/SATSU Workshop, March.

[33] Schaefer, E. (1991), 'Cetus retains PCR patents', *Nature,* 350, 7 March, p. 6.

[34] Schankerman, M. and Pakers, A. (1986), 'Estimates of the value of patent rights in European countries during the post-1950 period', *Economic Journal,* 96, pp. 1052–76.

[35] Sederoff, R. (1993), Letter to *Science,* 259, 12 March, pp. 1521–2.

[36] Spinks, A. (1980), *Biotechnology: Report of a Joint Working Party,* HMSO, London.

[37] Von Hippel, E. 'Co-operation between rivals: informal know-how trading', *Research Policy,* 16, pp. 291–302.

[38] Webster, A. (1994), 'University–corporate ties and the construction of research agendas', *Sociology,* 28, 1, pp. 123–42.

[39] Webster, A. (1994), 'UK government's White paper (1993): a critical commentary on measures of exploitation of scientific research', *Technology Analysis & Strategic Management,* 6, 2, pp. 189–201.

DEALING WITH BIG BUREAUCRACIES: PROBLEMS FOR THE SMALL TECHNOLOGY PRODUCER

JIM HANLON AND PHILIP L. GARDNER
TRIUMF Technologies Inc.
4004 Wesbrook Mall, Vancouver, Canada V4A 8M2

1. Introduction

In the late 1960s the three universities in the province of British Columbia persuaded the Federal and Provincial government to fund the building and operations of TRIUMF (TRI University Meson Facility) as the major Canadian national laboratory for research into sub-atomic physics. The first beam was produced from the 520 MeV cyclotron accelerator, the heart of the facility, in December 1974, and the laboratory has been operating ever since at its location in Vancouver, adjacent to the campus of the University of British Columbia. The three original universities have now been joined by five other Canadian universities from across the country, in the management and operation of TRIUMF, with the operating funds for the facility still provided by the federal government.

Although the primary mandate of TRIUMF continues to be the pursuit of excellence in research into sub-atomic physics, for the past ten years the Canadian government has been encouraging government-funded research laboratories to commercialise their science and technology knowledge. By 1990, TRIUMF had responded to this government pressure by establishing a one-person technology transfer office. TRIUMF itself has a total staff complement of about 300, but all of the commercialisation activities have had to be administered and managed by teams of about four people. Over the past five years, as a small technology producer, TRIUMF has had extensive dealings with big bureaucracies in the course of transferring technology out into the market place.

This chapter will review some of the issues and challenges that have been encountered by TRIUMF and other small technology developers in their dealings with big bureaucracies which, based mainly on our own experiences, are defined as both governments and large industry in North America, Asia and Europe. The approach that is being utilised here is to identify significant issues and, where possible, offer solutions that seem to have worked for us in the past. There is absolutely no presumption that we can identify all of the problems, and certainly we have no special claim to solutions. The object of this discussion is rather to lay out our experiences in dealing with large bureaucracies, and hope that we can provoke others to comment and critique.

There are numerous unexpected problems for the small technology producer dealing with the big bureaucracy, and this discussion does not pretend to identify, or supply solutions for, all of them. Doubtless, every small technology producer will encounter its

J. Kirkland (ed.), Barriers to International Technology Transfer, 65–75.

own unique problems, but hopefully the solutions that are reported here will suggest some possible approaches.

Following a brief review of the role of TRIUMF as a small technology producer, there are four main contact scenarios between the small technology producer and the big bureaucracy that will be discussed sequentially. They are: first contacts; negotiating a licence or contract; contract administration and liabilities.

2. The role of TRIUMF

TRIUMF is a research facility which has a primary mandate for research into sub-atomic physics. Its staff are almost entirely committed to performing research and/or the operation of the facility. It was not until 1990 that, under a specific financial contribution from the provincial government, TRIUMF was able to appoint one full-time person to be responsible for commercialisation activities. These activities are similar to those found in most commercialisation offices of small technology producers and include all aspects of patenting, identifying potentially commercial technologies, identifying potential commercial associates, negotiating terms for licences and contracts, plus the routine aspects of public relations, and financial and legal issues.

At any one time there are about thirty commercial projects requiring greater or lesser attention, being a mixture of licence negotiations, sales of specialised products manufactured by TRIUMF, and consulting contracts. TRIUMF has been assigned a number of patents by the researchers, in return for funding commercialisation activities, and holds ownership of all of the licences.

An ironic dichotomy has arisen as TRIUMF has become more successful in its commercialisation efforts; we now face an ever increasing administrative burden from the commercial activities. For the fiscal year 1995/6 it is projected that TRIUMF will receive about C$3 million in commercial revenue, which represents a substantial increase from the C$150,000 of six years ago. However, these larger amounts of money have tended to come from larger institutions, such as governments and big corporations. One attribute of these large bureaucratic organisations that became quickly apparent is that they have whole departments dedicated to contract negotiation, contract administration, and law and finance, which substantially increases the burden on the small group trying to transfer technology from TRIUMF. Now we are mired in patenting issues around the world, f.o.b. delivery points and all of the insurance and other aspects that go with it, together with ongoing negotiations that never seem to end, on numerous issues we had barely even heard of several years ago.

3. First Contacts

The first challenge facing any small technology producer is to establish a productive initial contact with a prospective commercialising big bureaucracy. Many large companies and institutions with their own research capability routinely refuse to even review an unsolicited technology innovation. The objective of this approach is to protect any potential claim to intellectual property rights on similar technology that they may be developing. While the corporate objective may appear reasonable, it completely blocks any chance of technology transfer from a small producer. This is particularly significant, since institutions that are working in closely related areas of research are the ones most likely to recognise the potential commercial market for some new, outside technology.

The most effective method to circumvent the corporate policy of refusing to review outside technology appears to be through establishing some personal contact at a senior level in the target institution, prior to introducing the technology. If no introduction can be arranged through a third party, we have found, at TRIUMF, that it is productive to first try and establish our credentials, by supplying documentation, such as annual reports and research reports, before the innovative technology itself is introduced. This has been particularly necessary in dealing with large organisations in Japan and Korea.

Even when the principle of access is established, locating the right contact within the company can be difficult. Transferring technology to a large company, or other bureaucratic institution, almost always requires an individual within the recipient organisation to recognise the *potential* in the market as well as the innovative technology. Invariably this means that the initial contact becomes critical. We have found this to be particularly crucial in dealings with the major trading houses and conglomerates of Japan and Korea.

In Europe and North America the first contact is more effective the higher it is in the bureaucratic system, from which point you will be directed to the appropriate contact, which will immediately provide an insight into your perceived importance. In Asia, however, although it is important to have some senior contact as a measure of respect, the technical expert will also expect a direct contact in recognition of their significance to your project. At TRIUMF we have found it essential to employ the services of a domestic contact to guide us through the local customs and systems.

3.1 TECHNOLOGY OWNERSHIP

The key to commercial activity in any research institution is clearly the scientists themselves, since they are the repository of the knowledge and understanding that represents the economic rent of the process, or product, which the small technology producer is marketing. The actual owner of the intellectual property may be either the individual researcher, or the small technology producing institution, according to established laws, terms of employment and policies. The scientific culture is quite distinct from that of the commercial world, however, since scientific researchers have a tradition of academic openness and exchange. To a tightly controlled bureaucracy this is perceived as a potentially leaking sieve for 'their' technology.

The basic rule adopted by TRIUMF is to assume that the rights to any specific work paid for by a commercial enterprise is owned by that enterprise, unless there is a specific disclaimer to the contrary. However, this does not mean that a contractor can retain TRIUMF for a C$10,000 contract to provide a study in an area of our expertise, and then claim rights to all our intellectual property in that area. The difficulties arise where a scientist is researching an area that is close to the technology that has been assigned to, or owned by, a commercial licensee or contractor.

The company management often believe that they have some restrictive rights to all of that technology, and that related material should not be published without their approval. As a research establishment, TRIUMF cannot jeopardise its academic freedom to investigate an entire area of discovery every time it enters into a commercial contract. Of course, this problem can arise with a company of any size, but with big bureaucracies, whether they are government or industrial, the problem can quickly escalate when they subsequently send in a team of managers and lawyers to solve the problem.

The right to restrict the academic and alternate commercial use of a technology that has received some limited outside funding is a particularly difficult issue to deal with, since it frequently produces strong emotions on both sides. The researcher believes that the company or government department is trying to wrongfully obtain or restrict research results, and the bureaucracy believes the research producer is threatening the value of the technology that they paid for. The optimal solution is to have a written understanding detailing the ownership of the technology that is signed by both parties before the licence or work contract is signed.

If a small technology producer does get into an after-the-fact disagreement over intellectual property rights, it can quickly devolve to major disagreement, if the big bureaucracy starts following some standard procedures and sends in a group of 'specialists' to solve the problem. These are frequently people one has never met before, and who are quite removed from the actual issues at hand. Before the situation mushrooms completely out of hand, it is necessary to cut through the bureaucracy to a senior representative who can help actually resolve the situation. If things do really reach an impasse, it may be necessary to employ the services of a professional arbitrator, or mediator, or, as a last resort, bring in a lawyer. However, at all times, as soon as the situation appears to be getting difficult, ongoing legal advice is essential.

3.2 PUBLICATION AND CONFIDENTIALITY

Many large bureaucratic organisations have difficulty dealing on major technical matters with small technology producers, because of the perceived risk of leaks of technical information. The most frequent source of potential leaks of commercial technology that TRIUMF has had to deal with has been from the researchers themselves providing confidential information to outside colleagues, or publishing it in a paper. This is the result of the mixing of academic and commercial cultures, and not through any deliberate action by the staff, although, of course, that cannot be excluded as a possibility.

The inadvertent release of information can be fairly simply controlled by making the researchers aware that once a technological innovation has been openly disseminated, through for example publishing, then that knowledge is freely available for anyone to utilise. The exception is in the United States, where the Patent Office works on the 'first to invent' basis, and allows patent filings for up to one year after public dissemination, but that does not apply in Europe and Asia.

To further avoid any costly mistakes, one or two persons within the TRIUMF team are designated as having authority for everything that is released on a specific project. This includes academic papers. Large bureaucratic organisations will frequently demand the right to review all papers that are proposed for publication in areas that touch on the topic of technology. We have found that this outside vetting of research publications must be resisted as strongly as possible, or it can quickly escalate to outside bureaucracies reviewing every paper that is produced at the facility.

As a corollary to this, TRIUMF has frequently experienced attempted incursions into its right to publish academic results. From discussions we have had with other universities, this is a fairly common problem. As a research facility owned by Canadian universities, with many of the research staff being faculty members at one or other of our member universities, any outside restriction on the freedom to publish is completely unacceptable.

TRIUMF has found that the issue of freedom to publish is one that must be clearly established at the very start of discussions with potential commercial and other bureaucratic institutions. Despite some strong pressures from their negotiating teams, we have usually managed to get even the largest bureaucracies to agree, *a priori*, that TRIUMF would retain academic freedom to publish, with the proviso that we would monitor our own publications, to ensure no confidential information is released. Where overriding government regulations have demanded previews of articles to be published by TRIUMF staff on a specific project, we have always obtained a reciprocal arrangement.

The concept and requirement for technical confidentiality has been explained to the researchers at TRIUMF and they have handled the commercial aspects of confidentiality with no problems. It was more difficult to convince the scientists not to freely exchange potentially commercial technical information with colleagues outside of their facility until it had been adequately protected. Although it was close on one or two occasions, there have been no major technology losses from TRIUMF in recent years.

3.3 THE ROLE OF PATENTING

Patenting can be useful in defining the precise parameters of the intellectual property implicit in the technology that is owned by the research institution and is being transferred. Big bureaucracies also respond well to the clear legal specification and claim that is incorporated in a patent. However, patents require a fixed process to be followed, they do not provide academic credits, they are expensive, and patents do not transfer technology. TRIUMF has been faced with the 'patent or publish' issue on a number of occasions.

For a small research facility, the optimal approach is to file for a patent at minimal cost, and then use that as the basis for negotiations on transferring the technology. Where possible, it is wise to arrange the timing of the patenting procedure so that the large bureaucracy pays for the later patent costs. If a researcher prefers to publish a paper, rather than patent an invention then, as long as he is aware of the options and issues, that is the researcher's choice. Of course, on several occasions we have filed a patent quickly, so that an article can be published immediately afterwards.

4. Negotiating a Licence or Contract

The most strategic factor that we at TRIUMF have observed in dealing with big bureaucracies is that there are many discrete groups within the bureaucracy, and each is charged to perform a specific function. Lawyers pursue legal issues, accountants pursue financial issues, and contract adminstrators administer contracts. In this type of large organisation there is virtually no scope for such specialists to stray beyond their defined function. The result is that when a small research facility has to deal with such compartmentalised activities, it can be very difficult to establish useful discussions of more global solutions involving more than one functional group, unless a senior decision maker is involved. The result can be that the small technology producer spends time rushing from department to department, acting as an information liaison facilitator, and consuming considerable amounts of precious time. To minimise this effort we have found that establishing effective contact points is an essential part of every negotiation with a large bureaucracy.

4.1 ENTRY POINTS INTO THE BUREAUCRACY

The experience at TRIUMF has been that the process of transferring technology to a big bureaucracy starts with the initial contact at the technical level. Initially, there is a flush of easy agreement and ready cooperation, and the corporate scientists return to their company to extol the virtues of the intellectual property that they have been shown. At this stage the formal negotiations are placed in the hands of the corporate contract team. Suddenly it can become a formidable experience for the small research institution, like TRIUMF, as we are faced with this new team, or teams of specialists.

There will be at least one lawyer, and frequently more: a supervising lawyer, a contracts lawyer, and an intellectual property lawyer. Then of course there is the corporate negotiating group of at least two persons: a supervisor of contracts, and a contract negotiator. Next of course the company has to have its business interests represented through a business manager. Finance will be represented at some stage through a supervising accountant, and a bookkeeper. Each of these parties has an integral, albeit independent, part to play in the contract or licence negotiating process, and has to be included for the agreement to be accepted in the bureaucracy.

What is almost certain when dealing with big bureaucracies employing multiple groups such as those listed above, is that communications between such groups is often quite limited. This means that information cannot be sent just to the company, but has to be directed to a specific group, or groups, under the assumption that it will not be forwarded to any other group. At TRIUMF we have found this to be true in our extensive dealings with both government bureaucracies and large corporate organisations. The only solution that we have been able to find is to have the entry points into the bureaucracy formally identified for each group, with the name of an individual.

4.2 FORMING A NEGOTIATING TEAM

The small technology producer is faced with an overwhelming task when trying to negotiate with a large bureaucracy. As pointed out in the previous paragraph, first there is the disparity in numbers and support staff, and possibly even more important is that the team from the large bureaucracy is likely to be negotiating contracts every day, and has been for years, or even decades. Most people like to think that they are skilled negotiators, but to go up against seasoned professionals with an inexperiencd team is simply foolish. In most cases, you have to live with the negotiated contract for a number of years, and the experience of a poor long-term contract can be quite chastening.

Lawyers are expensive, and as a general rule we use them minimally. However, if a contract is worth hundreds of thousands, or even millions of dollars during its lifetime, then funds spent on lawyers at the beginning can be money well spent. This can also be the one opportunity for the small technology producer to intimidate the big bureaucracy, with an equally skilled negotiator. The most important point to remember is that the lawyer works for you, and you must direct them, and keep on top of the negotiations yourself. A lawyer is your consultant, not your replacement.

4.3 IMPLICIT TERMS AND CONDITIONS

A major issue in dealing with both government and commercial bureaucracies is their reliance on established terms and conditions, that are not routinely revealed in any detail to the small technology producer that is planning to enter into a contract with them. In 1994, when TRIUMF starting negotiating for a contract with the United States federal government, we were provided with a list of about one hundred 'FARs' that applied to the contract. We employed a large, expensive law firm in Washington D.C., and gradually became educated that FAR stands for the numerous Federal Acquisition Regulations that can be found in four volumes published by the US governemnt. Our lawyers, who have been an invaluable investment, supplied us with copies of these FAR volumes, with advice on their applicability. It quickly transpired that only about half of the FARs that were originally going to be imposed on us were actually applicable.

The US government FARs are certainly not a unique example of this problem of big bureaucracies assuming a third party has knowledge of some standard set of complex terms and conditions that are only identified by initials and numbers. TRIUMF has experienced the same types of references to such standardised terms from both the Canadian federal government and large private companies. In each case, there has been an oblique reference to the applicability of a standard set of terms and conditions, and in none of the instances was the referenced material provided.

The only reasonable solution to this issue appears to retain the services of expert legal counsel, who are well versed in the particular standard terms and conditions, since some seemingly innocuous paragraphs can have potentially far reaching effects. Of course, the technology producer should make sure that one or more of its staff has copies of the relevant standard terms and conditions, and every attempt must be made to monitor and control this whole area.

4.4 MANAGING THE BUREAUCRACY

Given the number and discrete nature of the various groups that are responsible for individual aspects of the negotiations on behalf of a big bureaucracy, the paper flow can quickly become quite intimidating to the limited resources of a small technology producer. At TRIUMF we have watched in amazement as the paper has flooded as a daily tide from our large commercial and government collaborators. Because the big bureaucracy has dedicated resources specifically to handle this type of procedure, the sheer volume of paper can become a real obstacle to effective communications between the two parties. It is difficult to discuss issues if the other party keeps referencing information that you cannot even identify. Thc only solution to the bureaucratic paper flood is to have an effective filing system in place from the start. We have only been able to retain control of the paper when we have made a concerted effort to dedicate a person and adequate filing space to 'paper control'.

Every large bureaucracy that we have had to deal with has had established procedures that are followed rigidly internally, and that they insist must be followed by any small party that deals with them. Some of these bureaucratic procedures may be quite acceptable, but equally some may be unnecessarily onerous.

It obviously seems trite to suggest that the two parties simply negotiate acceptable procedures, but in our experience with big bureaucracies, this can be an essential but difficult process. Invariably the other negotiators initially dismiss the suggestion that any of their established procedures be changed or dropped as completely unworkable. However, we have found that, after prolonged and forceful discussions, changes can always be accommodated. The advice and original drive for challenges to irrelevant bureaucratic procedures originated several years ago with the legal counsel who was advising TRIUMF, and it has proved to be sound advice.

As a corollary to the previous issue relating to the established bureaucratic procedures, both governments and large corporations will normally have standardised terms and conditions that are completely slanted in their favour. This is quite understandable from their perspective, but we have found that many of the individual terms should be reciprocal in their effect on the two parties. This can apply particularly to such items as notification of impending contract changes and approvals for such things as press releases and publications. Again, the solution is simply to identify those contract items that should be reciprocal, and point out the reasonableness of that position. There can be some interesting discussion, but we have found that our reasoned argument usually prevails. If nothing else, it provides a bargaining chip for future trade-offs.

The very function of a bureaucracy is to compartmentalise operations and establish a chain of command for decision making. This can create quite a problem for the small technology producer if the result is inordinate delays in obtaining corporate approvals. The two areas that have caused problems for TRIUMF on several occasions are formal approval of contract terms and approval for payment of accounts. All questions that are raised about why it is taking so long are met with the standard bureaucratic response that it is 'in the system'. If the small technology producer is in a strong enough position, it can fix what is known in North America as 'Drop Dead' dates. This means that firm dates are established ahead of time such that if the bureaucracy has not delivered the approvals through the system by that date, the technology producer stops all work, and the contracts are considered terminated. If the position is not that strong then difficult decisions have to be made as to how far the small technology producer can proceed with the risk of not being paid, and/or losing some rights to the technology.

4.5 ESTABLISHING A VALUE FOR THE TECHNOLOGY

One of the most contentious issues between any technology producer and a technology user is the valuation of the technology. In financial terms, the objective of the two parties tends to be in direct conflict, with each trying to maximise their own return. In fact, it must be remembered that both parties will profit from success in commercialising the technology and both will lose from a failed commercialisation attempt. At issue is the division of risk and return.

The big bureaucracy may attempt to assign as much of the risk as possible to the smaller technology producer, for the absolute minimum cost. There are two basic financial elements that have to be considered: which party bears the costs of development; and what is an appropriate royalty return on the technology.

If there are several organisations that can be encouraged to compete for the rights to the technology, then the value can be set by the competitive market forces. In reality, if one of the contenders for the technology is a big bureaucracy, it is unlikely that value can be established through this method. The only alternative is defining a value for the technology based on the full cost of reproducing similar technology, and for the technology producer to then push toward that value with controlled brinkmanship. Generally, the only major lever that the small technology producer has is to impede, or deny the bureaucracy the right to exploit the technology. As has already been pointed out, that position is unlikely to benefit either party, but the threat may have to be subtly introduced to force an appropriate return for the technology. In general, the party that is projecting to realise the majority of the commercial returns should bear the majority of the costs of development, as an investment. Once the technology has been successfully produced to the commercial product stage, the technology producer should obtain a royalty return on the gross revenue, bearing in mind that if a commercial producer has a 10 per cent gross margin, then a 5 per cent royalty equates to a half ownership in the production.

5. Contract Administration

Once the contract has been successfully negotiated between the small technology producer and a big bureaucracy, it must be administered by both parties, usually for a significant period of time. Here again the difference in size between the two parties makes it very difficult to have a balanced arrangement. This is not necessarily because the larger entity wishes to take advantage of the other, but rather because of the administrative procedures and processes of a bureaucracy are invariably much more complex.

One example of this causes excessive workload requirements. A large bureaucracy invariably has a number of corporate departments that have some involvement in the administration of a contract. Most commonly these include accounting, contract administration, legal and audit, and operations. If each of these groups places ongoing demands on the small technology producer, then the latter will have at least one full-time person constantly administering that one contract, and consuming any potential profit.

Again, this issue should be resolved between the parties during contract negotiations. Most large companies and governments recognise that a small technology producer is not in a position to dedicate endless staff time to contract administration and will agree to a reasonable compromise. Provided that the small producer establishes an agreed set of auditable administrative procedures—for example, on a desk-top computer—that should meet the minimum requirements of the larger entity. If not, then contract administration must be a chargeable item for the technology producer within the contract.

As a corollary to the above issue, big bureaucracies tend to be bound by their established methods, procedures and formats for contract administration. For example, a large bureaucracy will generally have payment procedures that are completely inflexible. In some cases, at TRIUMF, we have encountered governments that routinely pay accounts up to six months after the invoice is submitted. This can prove quite troublesome for a small technology producer whose financial resources cannot stretch to accommodate such procedures. At TRIUMF we have been somewhat successful in overcoming this problem by making it a clear issue in negotations, and asking for a rotating security fund to be established at the start of the contract. That fund should be sufficient to cover an agreed number of projected outstanding invoices, and can be drawn down by TRIUMF after, say, thirty days of issuing an invoice.

5.1 REASONABLE OVERHEADS

Some level of overhead expenses will obviously have to be incurred by both parties, but the burden will probably be felt more acutely by the small technology producer than by the big bureaucracy. The previous two items have touched on this issue already, but there can be a number of overhead items that can become quite costly. Some of the more obvious are travel costs for regular meetings, as well as for technology support, and the various aspects of shipping costs.

Once again, when the agreement is negotiated, the smaller technology producer should endeavour to include as many items as possible as chargeable costs. These relatively innocuous items become issues because the large bureaucracy has well established costing methods that include allowance for such items, and simply overlooks the significance of such items to small entities. It is incumbent on the small entity to raise these issues early on, and the experience at TRIUMF is that accommodations can always be made.

6. Allocation of Liabilities

One of the most difficult issues to resolve adequately between small technology producers and large bureaucracies is the allocation of liability, inherent in the production of technological products. The large bureaucracy frequently perceives a liability threat from new technology because it is the larger, more established entity that tends to be held culpable if anything goes wrong. Fortunately it is an area in which the bureaucracy has experience, and can accept a measured amount of risk, but corporate and government policy invariably dictates that such risk must be minimised. Since there is a finite amount of risk, and only two parties, if the bureaucracy decreases its risk, the small entity must have its exposure increased.

6.1 ALLOCATION OF TECHNOLOGICAL LIABILITY

The types of liabilities that are considered here would result from such things as transportation of dangerous materials, failure of a technical innovation and other physical and financial threats resulting from the development and use of the technology. The preferred position for the small technology producer is to attempt to limit its liability to the provision of the technology, with no implied guarantees or representations attached. If this is not possible then the liabilities must be minimised by working in conjunction with insurers, lawyers and accountants.

7. Conclusion

This discussion can only touch briefly on the myriad of issues that have to be addressed when a small technology producer deals with a big bureaucracy. The underlying principle must always be remembered, that for an agreement to work between two parties, it must benefit both of them. No matter how exciting the technology, and how tempting an arrangement looks with a large established bureaucracy, if it does not provide an appropriate return at an acceptable level of risk for the small entity then it should not be pursued.

As was discussed at the start, TRIUMF is a joint operation of a number of universities in Canada, and we have been transferring technology quite successfully for several years. However, if the preceding discussion sounds as though we have all of the answers, that is definitely not the case. We have had our share of disasters in technology transfer, and there are a number of experiences that we would not care to repeat. If we are at all knowledgeable about these issues, it is because it is easier to learn from mistakes than successes.

MANUFACTURING TECHNOLOGY TRANSFER BETWEEN WEST AND EAST EUROPE

R.J. GRIEVE,* T. KOCH** and E.A. PEIRCE*
Department of Manufacturing and Engineering Systems, Brunel University, UK
**Institute of Mechanical Engineering and Automation, Technical University of Wroclaw, Poland.*

1. Introduction

The authors are practising engineers and members of Universities in the United Kingdom and Poland, and have been involved with technology transfer for a number of years. Involvement with the transfer process has been not only through the usual research and development activities between universities and industry in the individual countries, but also at an international level between their respective countries. Involvement in the United Kingdom has been with various research council initiatives to allow universities to form closer links with industry and work on projects of immediate industrial relevance. The significance of these has been the direct involvement of university academic staff with the company on a frequent and regular basis. The research staff on these projects have either worked in the company for the majority of their time or have been seconded from the company to work in the university laboratories for the duration of the project. This feature of close collaboration invariably results in successful projects and can also be used effectively in international activities. At the international level the technology transfer has taken place through involvement with European Union initiatives such as TEMPUS Programmes. A feature of a particular programme referred to here has been the exposure of Polish engineers and students to United Kingdom industry in addition to the transfer of the technology as defined in the tangible sense of machines and computer equipment. The comments that follow are therefore the result of practising the process of technology transfer. Although the comments are intended to be objective they will doubtless contain the biases and prejudices of the authors but the experience has been gained from the first hand experience of actually attempting to transfer technology. As might be expected from the above the comments will relate to the transfer process between universities, industry and countries. The authors' professional interests are concerned with manufacturing industry and the application of technology in the broadest sense. This covers both the implementation of technology in addition to its organisation and management once installed. This might be summarised as manufacturing technology, manufacturing methods and manufacturing strategy. All of these must be considered if the

J. Kirkland (ed.), Barriers to International Technology Transfer, 77–89.

transfer process into industry is to be successful. The following discusion is therefore centred around transfer into the manufacturing industrial sector of the technology and operational methods needed to enhance performance. Invariably the technology has been computer controlled equipment for use in manufacturing industry. There is no doubt that in all these activities it is the commitment and enthusiasm of the people involved that has led to their success rather than the degree of technological sophistication of the equipment. In other words people are more instrumental in success than the technology. It is the implementation that matters and not 'how good the technology is'. Technology can only perform well if correctly implemented and managed. This is a factor that is often forgotten when dealing with technology transfer. Far too much emphasis is placed on the technology and not enough on people.

2. Advanced Manufacturing Technology

According to the United States Office of Science and Technology Policy [14] manufacturing is one of the important areas for a country's economic development. It includes flexible and computer integrated manufacturing systems, production management systems and also manufacturing machinery which uses artificial intelligence. The growth of produced goods in both quantity and quality depends on a sound manufacturing infrastructure of manufacturing processes. Technology transfer in this area is therefore very important.

Many of the technology transfer activities are associated with what is described as Advanced Manufacturing Technology (AMT). By AMT in the manufacturing industry sector we mean a broad range of technologies that may be software related, such as information systems, or hardware related in the form of machine tools or robots. Often organisational changes will result in the introduction of the technology and this must also be considered and included within the broad definition. A useful definition when considering AMT is in relation to the technological changes that might accrue from its introduction [5]:

> '... as alterations in physical processes, materials or equipment, which has an effect on the way work is performed .... or on the efficiency or effectiveness of the enterprise'

Therefore technology transfer should address all the issues that will result from technological change.

The introduction of AMT is an attempt by most companies to achieve greater control over production processes, product quality and product costs. In effect this implies that AMT can have a significant influence on the manufacturing enterprise. Simple implementation is not enough. The company is likely to have to undergo structural changes to accommodate AMT. There are four important areas where the impact of AMT will be particularly felt by the workforce as follows:

- The individual: new skills, multiple skills, flexible working and continuous training.
- Control: flat and participative rather than procedural structures, local autonomy rather than rules, decentralised and networking.

- Culture: organic and flexible, shared value and common purpose.
- Work organisation: functional integration, matrix management and project teams, shared resources and de-specialism.

New technology is not a simple product that can be bought 'off the shelf', rather, as noted by Bessant [2]:

'it is a field of opportunities offering a wide range of choices to suit an equally wide range of users'.

There is a wealth of standard products available to manufacturing organisations, ranging from machine tools to software products, but identifying, choosing and realising benefits through incorporation into the overall company structure can prove difficult.

As the importance of these issues has been recognised so has, increasingly, the complexity of the issues which need to be addressed. Several commentators have questioned the long established notion that technology management can be seen as a three-stage linear process, involving discovery of new ideas, transfer into a industrial environment and diffusion into new products. It is increasingly recognised that technology management is an interactive process in which organisations and technology constantly influence one another. This is often referred to as a 'configuration model' and is particularly applicable to the current generation of complex and highly integrated manufacturing technology which represents systems rather than single elements of technology [4].

More recently, a number of studies have shown that performance does not necessarily improve with the introduction of new technology. Thus emphasis is shifting away from the notion that the acquisition and substitution of technology will necessarily lead to improvements. Rather, it is now recognised that it needs to be accompanied by appropriate management mechanisms, with an emphasis on integration.

The introduction of AMT is not easy. It often lacks the commitment of management. It has been noted [1] that:

'although the majority of users achieve some benefits, the system is not performing to its full capacity in 75% of cases.'

The correct transfer and implementation procedure is therefore extremely important and all of the technological, organisational and human issues must be addressed. This is particularly so in the transfer process between the countries of the authors, where the culture, working practices and general 'state' of some industries in the receiver country may not be appropriate to accommodate AMT. These issues must also be part of the technology transfer process if it is to have any chance of success.

3. Technology Transfer

Technology transfer is an ill defined term that may be interpreted in many different ways. We are assuming that technology exists somewhere and that by some process it is transferred somewhere else. The word technology may imply that only 'hard' tangible objects such as machine tools, robots or computer systems are to be transferred. A wider interpretation would include knowledge, insight, educational methods, management techniques and the vast range of 'tools' that allow the technology (as defined in the narrow sense) to be first identified, then acquired and implemented, and finally used in some way. The use of the technology is likely to be found in manufacturing industry where the object of acquisition is to improve profitability and competitiveness.

Where does the technology come from? Is it from commercial technology suppliers such as machine tool manufacturers? Is it from research and development establishments or the universities? Is the technology developed in, for example, universities, implemented directly in industry, or is there some intermediate process to transform it into a product acceptable to companies on say a production line where 100 per cent reliability is expected.

The drivers for technology transfer are complex. Is the transfer a result of industry pull or research/development push. In other words, are there examples of technology development in research laboratories that may be solutions looking for problems or products. Will this be a driver for technology transfer? It is unlikely that this will be the case since a more powerful driver will be the industrial need to improve competitiveness and will result in the pull mechanism operating. Manufacturing industry must respond to market needs and will attempt to organise its infrastructure accordingly. It could therefore be argued that the transfer process is driven by market needs rather than the push of developing technology. How does industry know what it needs in terms of technology? How can it be made aware of what is available? Is it in a position to take advantage of it once it has implemented it?

Technology in the form of machines or equipment is tangible. This should not be a problem for the transfer process. If it were simply a case of physically transporting technology from one place to another then the technology transfer process would be a straightforward exercise. Unfortunately the term technology relates not only to the hardware but also to the knowledge of the people who developed and proved it, in an industrial sense, before the transfer process and also the skills and competences of the people who will implement it. This knowledge has to be part of the technology 'package'. With-

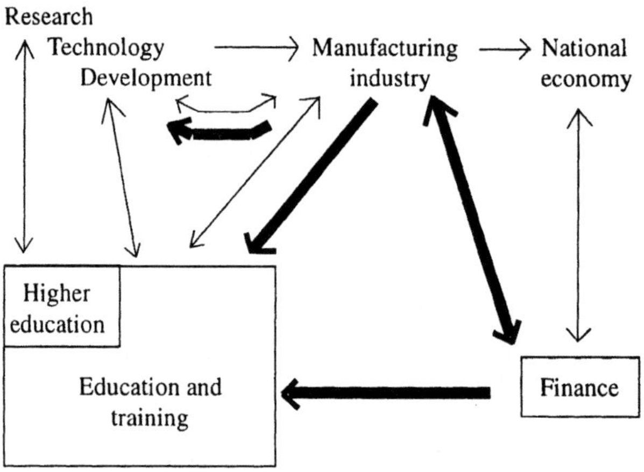

Figure 1. Manufacturing: flow of knowledge and finance [14]

out it the technology will at best be used in an inefficient inappropriate manner or at worse not used at all. If the knowledge, skills and competences are not available they must be 'obtained' from somewhere.This would imply that education is important and plays a significant part in the transfer process. It is also difficult to escape the need for finance in the transfer process. Someone has to provide funds for it to happen. This may be industry or governments. If a single economy is considered and the interrelationships between the various elements examined, a diagram such as that shown in Figure 1 gives a reasonable representation of the flows of finance and knowledge in relation to the role of manufacturing industry [14]. The fine double-headed arrows represent the flow of knowledge. This tends to be an intangible commodity but is probably best implemented by the movement of people and experienced engineers. The solid arrows represent the flow of finance through the system resulting in some effect on the national economy.

If a broader perspective is considered where technology transfer is between countries (in particular, in this context, between 'developed' western European countries and the eastern European countries) then a diagram such as that shown in Figure 2 might represent the interrelationships and dependencies. If countries are to take advantage of technology then the people and knowledge aspect cannot be overemphasised. Figure 2 is not intended to represent every conceivable link in the transfer process, rather it illustrates the important need to have links and dependencies between not only research and development but also education and manufacturing industry. Exposure to industry by the transfer receivers in the country from which the transfer is to take place is important so as to appreciate the current thinking in using and managing technology at the company level. Educational establishments can provide the catalyst for this to happen, particularly those

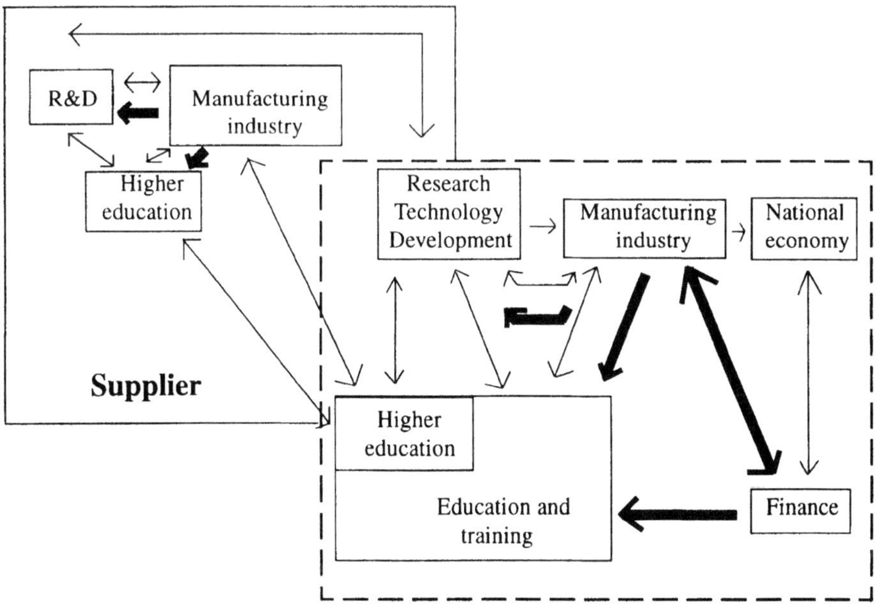

Figure 2. International manufacturing technology transfer process

that have strong links with industry at both the educational and research levels. Presumably time will reduce the influence of the technology/knowledge providers and the diagram will approach that shown in Figure 1. The general position in eastern Europe has been described by Koch [8] as:

> 'Central and east European economies, also the Polish one, experience a very difficult situation. Parallel to the integration process, they have to make up for the neglects of the totalitarian system. This process of transformation and reforms needs both technical development and a new organisational approach to the whole manufacturing process. The manufacturing capability of a manufacturing company essentially depends on manufacturing resources, manufacturing processes and production organisation. However education, qualification and motivation of human resources are the crucial factors for a company striving for the success.'

In eastern European countries such as Poland the production technologies and methods lag behind those of western Europe. This is not to say that the conceptual ideas for products is in any way inferior, it is simply that the process of concept to customer is protracted, and inefficient manufacturing methods prevail in many instances. This environment does not allow entry into highly competitive markets. Again Koch [8], being aware of this problem states:

> 'Awareness of this problem is at a preliminary stage, which allows consideration of not only the problems of industry, but also, first of all, the essence, methods and types of teaching.... Therefore we can submit the well known proposition that the development of a country's economy depends on its level of education.'

The awareness is that it is not only the technology that is important but the technological infrastructure and organisation. Education is seen as important with its gradual feeding of graduates into the system over a period of time. What is taught is obviously important and transfer of appropriate technology and methods into the educational system is crucial to future industrial success.

4. Technology Transfer in Poland

Government policy can influence the costs, access to capital, and investments in research and development. Government policy can also aid the process of technology acquisition by companies to allow them to be more commercially competitive. In the past some of the central and eastern European countries have suffered very little support for research and development spending. For example, in Poland in 1991 only $25 out of $4000 of GDP per capita was spent on research and development [13]. This has also been accompanied by a very weak pro-innovation policy. The Polish National Science Committee, for example, that governs the money for research and development in Poland, has given priority to basic, fundamental research in the past and has given less support to those projects which aim to implement new technology in the context of manufacturing.

Recently (in 1995), however, there have been some moves in Poland towards a pro-innovation policy [7]. There is a proposal at government level to establish an Agency for

Technology with the aim of facilitating the implementation of new techniques and technologies. Some changes have also been announced by the National Science Committee, who formulated the case for a pro-innovation Polish policy. The elements of these initiatives are, for example, modification of the finance–tax system and credit policy to support innovation in industry, an increase in the GDP quota spent on research and development, encouraging innovation and the creation of conditions and infrastructure to allow these initiatives to flourish. The speed with which these initiatives and policies are implemented is crucial for the development of Polish industry. At present there is little incentive to invest in new technologies and develop new products. In 1993, for example, only 1.6 per cent of sold products (in terms of cost) were new products [7].

5. Channels for International Technology Transfer in Poland

5.1 UNIVERSITIES

Universities can play a substantial role in technology transfer. Their activities could be categorised into four groups that are common to many countries:

- A primary objective is to provide the industrial companies with educated young people (BSc, MSc, PhD courses).
- Part-time study schemes (post diploma courses).
- Short courses and seminars.
- Research and development projects financed by industrial companies.

After political change, and above all the transformation towards a market economy, the technically-oriented university units have been trying to be more active in these areas. Because of the lack of substantial support at the national level (especially in the case of Poland) the stimulus for the new initiatives in that area often comes from international cooperation. The European initiative TEMPUS (Trans European Cooperation Scheme for University Studies), a part of the PHARE funding, has provided an important mechanism for updating undergraduate and postgraduate programmes at universities in central and eastern Europe. For example in the years 1990–94 the programme budget for Poland was 132.5 MECU and 348 projects were granted (for example, economy and management—73, engineering —86, protection of the environment —38) [12]. The TEMPUS projects are usually focused on the training of lecturers and students in European Union universities and on the updating of laboratories in the central and eastern European universities.

Examples of initiatives through the TEMPUS programme are the new study course in Automation and Robotics at the Mechanical Engineering Faculty of the Technical University of Wroclaw; the French MSc course on Production System Management; and the French model engineering college in Walbrzych, a branch of the Technical University of Wroclaw.

There are some other bilateral cooperation initiatives sponsored, in the case of the United Kingdom, by the British Council. The British Council has, for example, a programme for Polish industrialists which exposes them to high level management in overseas companies and to training in the United Kingdom (including training in United

Kingdom companies). Typical of other schemes (mainly supporting research) are: DAAD (Deutsche Akademische Austausch Dienst), Germany; Fullbright Advance Research Grants, USA; Obert Schuman Fund, France; Natural Science and Engineering Research Council (NSERC), Canada; Matsumae, Japan; Stefan Batory Fund, UK; Kingdom of Norway Mobility Programme, Norway; Bertha von Suttner—PhD scholarship, Austria; Universida de Murcia, Spain. Some of these programmes involve only a very limited number of grants.

In the field of part-time study schemes, cooperation based on international funding has produced a great impact. Good examples are the opening of one-year part-time study schemes in such fields as business, communication and quality management. The funding comes, for example, from USAID (United States Agency for International Development), or the European Union, in schemes such as TEMPUS or TESSA (Training and Education in Strategically Significant Areas). An important role in these courses is played by foreign lecturers coming from western universities. For example in a recent TESSA programme 40 participants, from middle and top level management in several manufacturing companies, were trained at the Technical University of Wroclaw in quality management [8].

USAID and European programmes like TEMPUS also play important roles in establishing special university units associated with centres of excellence, continuing education centres and technology transfer centres. Initially the courses offered by these establishments are more often related to business, finance, management and communication in an attempt to overcome the lack of development in these areas in the past. However the number of the courses that directly relate to technology is expected to grow.

At the Technical University of Wroclaw there is a newly created Continuing Education Centre supported by USAID, and a Technology Transfer Centre supported by the European Union TEMPUS programme. The Wroclaw Technology Transfer Centre initiative is supported by Brunel University in London and the University of Stuttgart (through Professor Pritchow and the Institute für Steuerungstechnik) in Germany. The areas of interest covered at the centre are: technology transfer to industry, innovation, realisation of joint ventures, consultancy, training of engineering staff, international cooperation, participation in student education and research and development. This initiative meets the needs and expectations of the Wroclaw local industrial sector. Senior managers of several companies have already formed a club interested in the initiative [8].

The UNIDO (United Nations Industrial Development Organisation) is attempting to implement a new initiative concerned with stimulating the growth of science and technology parks in central and eastern Europe.

The Copernicus programme, launched in 1994 by the European Union, provides a mechanism for international cooperation between universities from west and east and industrial partners in such areas as information technology, communication technology, manufacturing, measurement, agro- and food industries and biotechnology. All these activities have a positive impact on each other.

5.2 COMPANIES ESTABLISHED TO PROVIDE COURSES AND COUNSELLING

There have been a number of companies focusing on developing and running educational courses and providing consultancy. They usually have the cooperation of foreign companies and organisations. Examples in Poland are: ZETOM, a company active in the field of quality with a German partner; TUV Brandenburg; and OTREK, active in business and management with partners in Germany, Austria, Denmark and France [6]. Such companies often hire professional staff for lecturing and consultancy from universities.

5.3 FOREIGN INVESTORS

Technology can be directly transferred to the plant as a 'turn key' system through the involvement and investment of foreign companies. The good examples are ABB, Fiat, Beloit, Thompson, Phillips, and GM making investments in Poland

5.4 PRODUCT SUPPLIERS TO OTHER COUNTRIES

The companies in Poland that make products for sale abroad tend to cooperate with the companies they are supplying. The foreign companies 'force' the companies to develop best practices and use new technology and often train their staff. An example here might be Volvo and WSM Krotoszyn (a supplier of engine cylinder sleeves). This cooperation is beneficial to both companies, since the producer uprates the manufacturing environment whilst the buyer gains a high quality, competitively priced product. In effect this means both sides win.

5.6 OTHER INITIATIVES

There are other channels for international direct or indirect technology transfer such as:
- Local Chambers of Commerce which usually have contacts with those in other countries, for example, Lower Silesia Chamber of Commerce [6].
- Exhibitions and fairs.
- Associations such as the Club Polish Forum ISO 9000 which has approximately 400 companies as members whose interest is in best practice in quality management.

6. Barriers for International Technology Transfer

The following relates to the experiences of the authors through the technology transfer process applied to universities where the funding is usually through the European Union.

All projects tend to take too long to come to fruition. There is the time period between thinking about it, writing the proposal, bringing together a group of collaborators, submitting a proposal and waiting for the outcome. A recent Copernicus programme at the University of Wroclaw took 1½ years from conception to the granting of funds.

Polish academics/industrialists have not had previous experience of the type of activities found in the United Kingdom and elsewhere, where links between universities and industry are common. In the United Kingdom Technology Transfer initiatives have been in existence for many years. There are also advice centres that can supply advice on, for example, the DTI Guide to European Union Programmes and so on. Centres such as these would be useful for industry in Poland.

There is no real history of Polish academics working in Polish industry. Academic and research personnel tend to have had no experience of industry or working in industry, therefore this can present problems for the transfer process. In the United Kingdom many academics have had experience of industry and in Germany it is expected that the heads of institutes should have come from industry. This obviously can provide a better framework for the interaction of university with industry. There is also no system in Poland for pulling researchers into industry after completing a PhD programme. In Germany this is not the case and again this will help with the interaction process.

There are examples of companies who have had outside influences from foreign firms through investment and 'forced' manufacturing changes. Here the changes have been rapid. However if a company is totally Polish then the change can take a long time. The internal structures of the company will often need to be changed to accommodate the new industrial environment. Functional flexibility will be required with the workforce becoming multi-skilled, requiring training at technician and other levels. As systems become more complex and interdependent, as in modern manufacturing, and the environment becomes more turbulent and less predictable, so there will be an increasing need for flexibility. There will be a requirement for rapid response which will mean a focus on modes of work organisation that permit a degree of local autonomy and self organisation. In effect there will be a move away from a system that emphasises function or line arrangements and rigid control. Environmental uncertainty is a driver towards functional integration. Greater uncertainty needs closer coupling between functions to be used to reduce time needed to respond. Loose independent structures take time to operate [3].

To implement new technology requires long timescales and is not simply a process of deciding what technology to buy and where to implement it. It is much more a process involving the future manufacturing strategy of the company. This may be a problem in eastern Europe where a manufacturing strategy will have been associated with the supply of goods predominantly to other eastern European countries. The inclination to incorporate technology in a substitutional mode (that is, simply replacing old obsolete equipment with the equivalent new technology) should be avoided where there is little thought of strategic objectives. To simply apply technology within the existing company structure where this type of advanced equipment has not been used before does not often lead to success. In fact there are many examples of technology being introduced in a substitutional mode where improvements in productivity within the localised area of the application in the company have been impressive but where the overall productivity of the company has not increased [2].

Many of the barriers to technology transfer are the same as those that exist in western Europe and often relate to the views of people making decisions. Bessant [2], using the Rogers [11] theory of innovation substitution and adoption, gives five areas that can affect the decision to adopt new technology. There must be some notion of the relative advantage of the new technology over what it is to replace. The degree to which the technology innovation is perceived to be better than what it is to replace is important. Perception is implementation-independent and what matters is that the person making the decision believes the new technology to be better. Complexity is also a key issue; it is the degree to which it is understandable to the user that is important. If technology is simply being substituted for an existing system, then the principles will be well understood and not appear complex. Alternatively, if the new methods are technologically advanced they will be more difficult to accept. Observability will make the adoption of advanced technology a more comfortable decision to make. If improvements can be seen by observing the new technology at demonstration sites or other companies where it has been successfully implemented, then the risk may be seen to be reduced. This argument supports the role of Technology Transfer Centres where equipment can be observed. The ability to make trials will allow the proposed user to use the new technology and again fits well with the idea of Technology Transfer Centres. Finally compatibility relates to the ease with which the changes fit in with the existing structures. It is easier to substitute technology into existing structures and this often does not encourage the idea of looking at the whole company structure before considering changes.

The notions of relative advantage, complexity, observability, 'trialability' and compatibility all rely on the interpretations of people. In western Europe advanced technology has been adopted in many industries for many years and to some extent the benefits are accepted. In countries where the implementation is more recent the acceptance of the benefits may be more difficult.

It has been stated that the only barriers to technology transfer are people and that everything else is fairly easy [10]. There is more than an element of truth in this statement. Technology transfer involves people both during the act of transfer and subsequently during the implementation. People on both sides of the transfer process will have their own agenda, prejudices, biases and requirements of the process. Education is therefore extremely important and is a key issue when deciding how best to make industry aware of the advantages of advanced manufacturing methods. The idea of forming centres where industry can observe technology and learn is attractive and it is in this area that exchanges between the universities of east and west can help. Commitment to the technology transfer process must be present within the universities, both at the organisational and individual levels, and needless to say regular meetings and communication help in this regard.

It is also preferable to have extended visits by personnel from both countries. It is particularly relevant in the manufacturing area that time be spent in the industry of the donor country to understand the managerial and organisational issues of technology implementation and management. Industry in the receiver country must also be involved and a useful mechanism for this is the link with the university. If in time Technology Transfer Centres are created in the receiver country then the development of the local industry should improve. Many of these issues have been addressed in the TEMPUS programme between Brunel University, the University of Stuttgart and the University of Wroclaw. A particular feature of the programme has been not only the transfer of technology to Wroclaw to create a new laboratory for computer integrated manufacturing but also the training of academic staff at Brunel and Stuttgart. A unique feature has been the extended periods of time spent in United Kingdom industry by Polish students and academics gaining first-hand industrial knowledge and experience. The programme has now been extended to include the creation of a Technology Transfer Centre which has a similar structure and objectives (see Figure 3) to centres in western European countries.

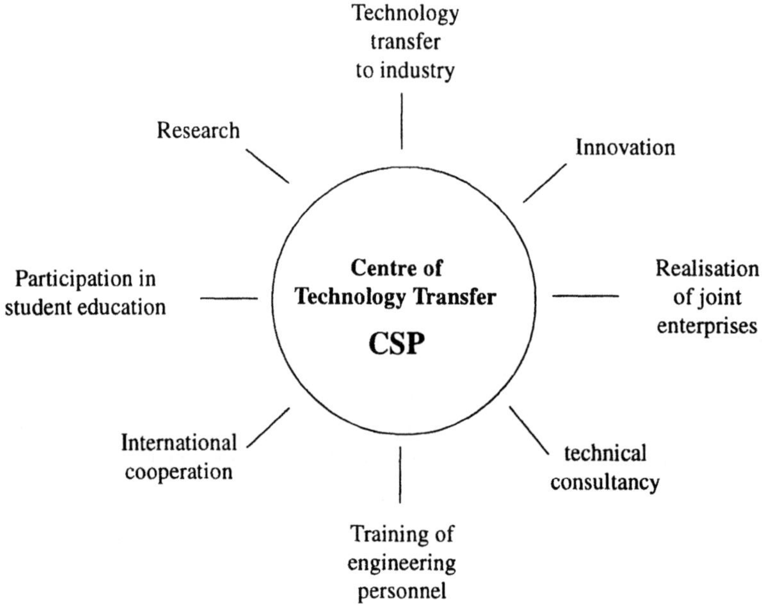

Figure 3. Aims and objectives – Wroclaw TTC [7]

Thus the objective is to address some of the requirements expounded earlier in the chapter for technology transfer in the manufacturing sector. These are to involve people in the transfer process at all levels and to consider also the organisation and management issues involved with manufacturing technology. A more complete description of the actual programme has been given by Koch [8].

7. Concluding Remarks

To say that there are no technological problems in western Europe is obviously not true and changes are often difficult to implement. However [8]:

> 'The industry of central and eastern European countries is in a more complex and difficult situation. Companies are worse equipped and sometimes they have still been centrally managed. They have also lost their markets and have major capital problems. The situation is often getting worse, because the personnel in the companies are not prepared for changes in manufacturing techniques, nor for changes in approach to work generally. This is a problem of mind turning.'

The experiences of the authors of this chapter, through their involvement with technology transfer in the manufacturing sector, believe that it is the total package of technology, knowledge, implementation and management skills that needs to be transferred. The industry into which the technology is to be transferred needs to be ready to receive it and will undoubtably need to change its structure and methods of working to take advantage of it. It is in this latter area that the role to be played by educational establishments in the receiver country can be extremely important through the development of courses for industry and Technology Transfer Centres. The technology tends to be the least of the problems, it is the people that are important.

References

[1] Bessant, J. (1985), 'The integration barrier ; problems in the implementation of advanced manufacturing technology', *Robotica*, 3, 97–103.

[2] Bessant, J. (1991), *Managing Technological Change*, Blackwell, London.

[3] Child, J. (1987), 'Organisation design and AMT', in Wall, T., Clegg, C. and Kemp, N. (eds), *The Human Side of Advanced Manufacturing Technology*, John Wiley.

[4] Fleck, K., 'The development of information integration: beyond CIM?', PICT Working Paper 9, Edinburgh University.

[5] Government of Australia (1980), *Report of the Government Inquiry into Technological Change*.

[6] *Investor's Guide Wroclaw—City and Province* (1993), Bogart Ltd., Wroclaw.

[7] Kastory, B. (1995), 'Cena mysli' (Cost of ideas), *Wprost*, 22, 28 May.

[8] Koch, J. (1995), 'Co-operation opportunities between east and west in education, research and development in the field of manufacturing technologies', Conference on Integration in Manufacturing (IiM): Opening Productive Partnerships—Concerted Efforts for Europe. Vienna 13–15 September.

[9] Myerson, J. (1995), 'Report focuses on legal barriers to business', *Warsaw Business Journal*, 1, 23, 1&9.

[10] Robinson, G. (1995), 'Technology transfer, a commercial chimera', ESRC Seminar Series: Barriers to Technology Transfer, CRICT, Brunel University, UK.

[11] Rogers, E. (1984), *Diffusion of Innovations*, Free Press, New York.

[12] TEMPUS (1995), 'Informator o projektach wspolpracy miedzyuczelnianej JEP: Nauki techniczne, czesc 1: Katalog projektow' (Guide for University Cooperation Projects JEP: Engineering Sciences, Part 1: Project Catalogue), TEMPUS Office, Warsaw.

[13] UNESCO (1991), *Statistical Yearbook*.

[14] United States Office of Science and Technology Policy (1991), *National Critical Technologies*, Government Report,, Washington DC: The White House, March.

[15] Yates, I. (1992), 'An industrialist's overview of manufacturing and economic growth', in Yates, I. (ed.), *Innovation Investment and Survival of the UK Economy*, Royal Academy of Engineering, London.

SOME ISSUES OF TECHNO-ECONOMIC DEVELOPMENT IN TURKEY

NAMIK K. PAK and ERGUN TÜRKCAN
The Scientific and Technical Research Council of Turkey
Ankara

1. Stages in the Techno-economic Development of Turkey

In 1923, when Turkey became a republic following the abolition of the Ottoman Sultanate, both her economy and her society (a population of 13 million) were largely agrarian. The first important step towards industrialisation was taken in 1934 with the introduction of the 'First Five Year Industrialisation Plan', when many industrial projects were set up with technical and financial help from the west and, more particularly, from the Soviet Union. These projects included the first Turkish iron-steel complex, several textile mills, sugar refineries, chemical plants, glassworks and even an aeroplane factory. At the same time many students were sent abroad for technical training and education. This then was the first step in the transfer of modern industrial technology to Turkey.

A second industrialisation plan was also formulated but had to be abandoned when outside forces, culminating in the second world war, dominated events. During this so-called 'Etatist' period nearly all industrial enterprises were owned and run by the state, there being a lack of private capital.

Significant political and economic developments took place after the war. No longer neutral, Turkey allied herself with the west and changed from single- to multi-party democracy. She was included in the framework of the Marshall Plan and in the declaration of the Truman Doctrine, drawn up to counter the threat of Soviet global expansion. Subsequent aid changed the face of technology in both civil and military life. Crucial to this so-called 'liberal' period was the birth of a new entrepreneurial business class. In addition agriculture became mechanised and modern infrastructures put into place with the creation of highways, harbours, hydraulic works, energy plants and so on. As an example of the rapid advance of mechanisation the number of tractors in use in 1948 was 1,756 and this rose to 24,000 in 1951 and then to 44,000 in 1957. Similarly, there were 268 industrial combines in use in 1948 and 6,592 ten years later. All of these were imports. For the first time the Turkish people were encountering machinery en masse.

In 1954 the 'Act on the Promotion of Foreign Capital' brought about the facilitation and regulation of foreign direct investment. This act still provides the main channel for private technology transfer and each of its amendments has liberalised it further.

With the early 1960s Turkey entered an era of planned economy but, unlike others, the Turkish five-year plans aimed at the accumulation of private capital and the creation

J. Kirkland (ed.), Barriers to International Technology Transfer, 91–98.

of a new class of industrialists. This did not necessitate the abolition of the state economic enterprises (SEEs), but their importance in the overall economy waned despite the increase in their absolute capacity. SEEs continued to be the main importer of important production technologies, particularly in the intermediate goods sector. In fact new industries and industrial entrepreneurs came into being during the planned era and established the backbone of Turkish industry. Turkey's main exports—of consumer goods, intermediate goods, consumer durables and the products of the automotive industry—all originate from this period. Thus also all the production technologies associated with these industries were transferred to the country by private national, multi-national or joint venture companies.

Many important goals in the industrialisation and modernisation of the economy were achieved throughout the planned era until, in the 1980s, a new shift occurred towards a more open market economy. The accumulation of private capital, brought about by competitive, outward-looking economic policies, had by then reached such a level that more emphasis had to be given to private investment, especially in the export and services sectors.

During this period of 'mostly-market-oriented' economy the plans continued to be implemented and at present the seventh five-year plan has been announced and is to be launched this year (1996). Subsidies, however, have continued to be a feature of the plans as they had been in the past, but since the 1980s there has been a gradual shift from the public to the private sector, particularly with regard to international dealings.

This period of liberalisation within the old economic boundaries is now coming to an end and emphasis is shifting towards a more global approach to techno-economic development. There are two reasons for this—the Uruguay Round or new WTO rules in the final act of 1995, and the Customs Union with the EU—both of which are important in setting guidelines for production and for the transfer of foreign technology. The Customs Union is the culmination of 33 years of preparation beginning with the Ankara Agreement of 1963.

If Turkey joins the EU in the foreseeable future the age-old dream of integration with European civilisation and technology, which started with the Tanzimat decree in 1839, will be fulfilled. For nearly 150 years Turkey has been acquiring a substantial amount of foreign technology— though not as systematically and intensively as Japan after the Meiji restoration of 1868. After the resistance to modernisation encountered in the 17th, 18th and early 19th centuries, barriers to European technology have now been almost put aside. In the last 409 years systematic technological transfer has been taking place and in 1995, at the end of the 6th Five-year plan, GNP distribution in Turkey was split between agriculture, 12.4 per cent, industry, 41.8 per cent and services, 45.8 per cent.

2. Methods of Technological Transfer

Nearly 30 years ago the methods of technological transfer to developing countries were set out in an UNCTAD paper [1]:

(i) the flow of books, journals and other published information;

(ii) the movement of people between countries, including immigration, return of emigrants, study visits and other travel;

(iii) foreign investment and the associated transfer of know-how and equipment;

(iv) import of machinery and equipment;

(v) technical cooperation programmes (multilateral and bilateral, official and private);

(vi) licensing, patent and know-how agreements.

In (i)–(v) above the flow of quantities, that is number of experts, periodicals, students and so on, can be measured in monetary terms but it is very difficult to quantify the technologies embodied in them. For this reason the sixth method, 'licensing, patent and know-how agreements', is the main source of measurement for technological transfer, as well as for producing 'technological balance-of-payments' tables. This type of transfer generally covers disembodied technologies, which are supposed to reach normal or near-normal pricing under international market conditions and even under imperfect market conditions.

Each of the methods outlined above are subject to their own barriers to technological transfer. For example, a technology agreement is subject to the seller's market which may have its own financial restraints. Another problem may arise with the age of the technology, which is closely related to the power of the buyer or the degree of monopoly of the seller. A buyer of technology may not necessarily find the latest, the second best or even an appropriate technology to meet his needs. Researchers in Turkey are concerned with all these problems but have not sufficient data to make more than a crude assessment of the relative importance of the transfer methods, values and the financial burden on the economy.

TABLE 1. Number of licence agreements by sector (1980–1992)

Year	Manufacturing	Agriculture	Mining	Services	Total
1980	10	-	-	-	10
1981	21	-	-	-	21
1982	25	-	-	-	25
1983	54	1	-	-	55
1984	59	4	-	6	69
1985	29	4	-	-	33
1986	68	3	-	5	76
1987	81	2	1	10	94
1988	68	3	-	5	76
1989	56	3	-	15	74
1990	62	2	-	6	70
1991	46	2	2	6	56
1992	42	1	2	3	48
Total	621	25	5	56	707
Per cent	87.8	3.5	0.7	8.0	100.0

Source: Office of Foreign Capital, The Treasury.

TABLE 2. Number of agreements in the manufacturing sector (1980–1992)

Manufacturing sector	1980-1992 Number	Per cent
Industries mostly producing consumable goods	45	7.2
Industries mostly producing intermediate goods	168	27.1
Industries mostly producing investment goods	391	63.0
Undefined/unknown	17	2.7
Total	621	100.0

Source: Office ot Foreign Capital, The Treasury.

With regard to the considerations outlined above, the main channel of technological transfer must be the patent licence and know-how agreements signed according to the 'Act on the Promotion of Foreign Capital, no. 6224'. Tables 1–3 give examples of the distribution of approved technology agreements, by sector and by donor country, signed under the Act between 1980 and 1992.

Out of a total of 707 agreements, 621, or 88 per cent, concern the manufacturing industry. Of these 391 were signed by the producers of investment goods, while the intermediate and the consumer goods producers signed 168 and 49 respectively (table 1). Within manufacturing, 102 agreements were in the automotive industry, 80 in electrical appliances, 77 in the non-electrical tools and machinery sectors, and 75 in the chemicals industry. 25 agreements have been signed for the agricultural tools sector and 5 involve mining (table 2). Of the 56 agreements in the service sector, 22 pertain to hotel management, a sector which increased particularly during the second half of the 1980s.

About 44 per cent of the licence agreements signed between 1980 and 1992 were signed with (former) West Germany (187) and the USA (120). The United Kingdom, France, Italy and Switzerland collectively accounted for another 30 per cent. The EU provides Turkey's main source of technology with 57 per cent of the agreements, and the USA around one third of it with 17 per cent (table 3). There is no specific data on either the payments arising from these agreements and their royalties, or any other characteristics. It is also very difficult to assess any financial burdens which may have arisen, as there is no reporting system in international banking dealings.

Law no. 6224 is not the sole channel for technology transfer to the country, as mentioned above. Domestic firms and SEEs can arrange technology agreements freely without any registration and, since 1980, can pay the royalties and other remunerations under a liberal exchange regime. It is therefore difficult to assess the characteristics of technology agreements signed other than under Law no. 6224. (Many bilateral and multilateral technological assistance agreements should actually be added to this stock of technology transfer agreements.)

According to its Central Bank, Turkey's overall burden of technology transfer between 1988 and 1992 was about US$770, of which 60 per cent was for technical assistance. This figure includes some visibles like books, films and so on (table 4).

TABLE 3. Distribution of licence agreements made under law no. 6224 in Turkey, by country (1980–1992)

	1980	1981	1982	1983	1984	1985	1986	1987
USA	3	3	2	9	7	1	8	19
Germany	2	7	11	15	28	5	26	23
Austria			1	2	3		2	4
Belgium				1	2			3
France		3	2	2	2	2	6	7
Holland	1	1		2	4	2	2	2
UK	1	2	1	3	6	2	9	7
Spain							1	
Switzerland	1		2	6	5	5		8
Italy	2	2	2	7	4	2	8	4
Japan			2		2	2	4	5
Others		3	2	8	6	12	10	12
Total	10	21	25	56	69	33	76	94

	1988	1989	1990	1991	1992	Total	Per cent
USA	15	15	12	19	7	120	16.9
Germany	20	15	20	8	7	187	26.4
Austria	3	3	1	1	1	21	3.0
Belgium	3	2		3		14	1.9
France	3	6	8	6	7	54	7.6
Holland	2	5	4	6	3	34	4.8
UK	5	9	3	6	6	60	8.5
Spain		1	1	1		4	0.6
Switzerland	3	9	4	2	4	49	6.9
Italy	3	2	8	3	4	51	7.2
Japan	5	1	1		4	26	3.7
Others	14	6	8	1	5	84	12.5
Total	76	74	70	56	48	707	100.0

Source: Office of Foreign Capital, The Treasury.

TABLE 4. Burden of transfer of technology to Turkey (1988–1993)

Items	1988	1989	1990	1991	1992	1993*
Technical assistance	47.7	56.5	102.5	115.4	141.3	125.0
Copyrights	1.3	8.9	5.7	10.6	12.3	13.4
Industrial intellectual property rights	14.4	18.5	27.7	19.2	28.3	24.9
Licences and trademarks	5.1	9.5	7.5	14.2	17.7	8.0
Computer leasing	17.7	13.6	10.8	11.0	9.3	19.7
Machinery leasing (excl. computers)	7.3	8.6	12.7	7.1	8.5	6.6
Total	93.5	115.6	166.9	177.5	217.4	197.6

Source: Central Bank of Turkey.

*Because the firms used several sources for their innovations,the sum of the percentages exceeds 100. 798 firms out of 834 have answered the questions about their sources of technology.

3. Some Features of Technological Innovation in the Turkish Manufacturing Industry in Relation to Technology Transfer

A recent survey on innovation in the Turkish manufacturing industry, covering 1,297 firms employing more than ten workers, showed that two-thirds of those firms changed their technology or introduced innovations in the five-year period between 1989 and 1993. There are ambiguities in the replies to the questionnaires, in spite of the standard definitions given in them, possibly because this is the first survey to be conducted in this area [2]. The concept of innovation in the survey did not strictly correspond to the Schumpeterian definition, or to the latest OECD definitions [3]. What it does mean is the renovation or replenishment of existing technologies in firms. The high percentage of intra-mural R&D activities as a source of innovation (64 per cent), together with other factors, must be interpreted within this context.

In fact, so-called R&D activity in innovative firms must be taken as complementary to the absorption process of a 'new' technology—new, that is, in the sense of new to the firm (table 5).

Of the 798 firms which changed their technology during the period mentioned above, 64 per cent are dependent on their own R&D activity as the source of technology. However, because the concept of R&D is so ambiguous in Turkey, this far from reflects reality. The main source of technology for firms seems to be the purchase of new investment goods, intermediate goods and raw materials (embodied technology), making up 54 per cent on average. Disembodied technology transfer—patent, licence and know-how agreements—makes up 29 per cent, reverse-engineering, 26 per cent and technical consultancy and assistance, 22 per cent of the total.

In the survey no indicator could be found to discriminate between the technologies transferred from abroad and those from domestic sources. But it is safe to assume that most of these embodied and disembodied technologies were transferred from abroad with no substantial R&D expenditure by the Turkish business sector (table 6).

TABLE 5. The sources of technological innovation in the Turkish manufacturing industry (per cent)

Sources of technology	Small firms	Medium firms	Big firms	Total
Firms' own R&D	62.27	65.31	67.52	64.16
Investment goods, intermediate goods and raw materials	53.75	54.76	54.70	54.26
Patent, licence and know-how agreements	17.05	37.41	48.72	29.20
Technical consultancy and assistance	19.38	23.47	29.21	22.43
Copying, reverse-engineering and publications	23.77	27.55	30.77	26.19
Firms' own patents	22.22	17.01	11.97	18.80
Others	13.43	13.94	18.22	14.91

Source: *Survey of Technological Innovation*, p.58, table 21.

TABLE 6. R&D expenditure by sectors in purchasing power parity (PPP)

	R&D expenditure (million $)					
	1983		1990		1992	
Sectors	PPP	Per cent	PPP	Per cent	PPP	Per cent
Higher education sector	133.5	56	597.2	70	1026.3	68
Public sector	68.3	28	84.2	10	124.4	8
Business enterprise sector	39.1	16	174.2	20	363.8	24
Total	240.9	100	855.6	100	1514.5	100

Source: State Statistical Institute, *Survey on R&D*.

4. A new Approach to Technological Development in Turkey. Enhancement of Technology Transfer and Production

The barriers to technology transfer in Turkey are mostly on the supply side. In other words, if there were a problem it would stem from the sellers, not from the purchasers of technology. Supply side problems generally arise because of the difference in the financial capacities of the buyers and sellers, that is, poor buyers and rich sellers. None of the projects which brought new technology were rejected on the grounds that the technology was inappropriate or out-of-date. Major evidence for the strong rate of technology transfer is given by the high growth rates of medium technology-intensive sectors of manufacturing, such as the automotive industry, consumer goods, consumer electronics. In high technology areas, however, such as telecommunications, defence electronics, aviation and semi-conductors, some obstruction has been felt. (Some examples of this are given by the problems faced by Turkish Avionics Industry in the production, under licence, of some of the electronics systems of F-16 fighters. The US government also permitted the transfer of second generation night vision goggles but blocked the transfer of the third generation. As Turkey is a member of NATO, this was not a restriction of the COCOM type. In the civilian nuclear energy sector, as yet not in existence in Turkey, the major powers call the tune in all matters of international technology transfer.)

The Turkish pharmaceutical industry is a major problem area in technology transfer. The difficulty lies with intellectual property rights and the old Ottoman Patent Law of 1879 which forbade the payment of royalties for any medical formulae for humanitarian reasons. Leading international pharmaceutical firms, particularly in the EU, demand the enactment of patent law as a precondition for entering the Customs Union and therefore made it clear that they intended to discontinue the licensing of new medicine patents to Turkey if a new law did not soon come into effect. However Turkish pharmaceutical firms argue that this would cause important price increases in the sector.

With the decision of the European Parliament pending, a new patent law came into force on 27 June 1995, and a reformed Turkish Patent Institute was set up. The new law postpones patent registrations for medical products until 2005 and for medical processes until 2000.

Some European manufacturers have already cancelled their licence agreements with leading Turkish manufacturing firms (mostly those dealing with consumer durables) in order to protect themselves against future Turkish competition inside the Customs Union. To combat this, Turkish manufacturing firms have begun to create a new R&D infrastructure with the aid of incentives from government, the most important of which involves a new technical regulation concerning the financial subsidies to be granted for domestic R&D activities. The regulation opens the way for subsidy of research activity conducted by firms or higher education/research establishments in line with Article 8 of the Agreement on Subsidies and Counterveiling Measures in the Marakesh Final Act of the Uruguay Round. The mission was given primarily to TÜBITAK in collaboration with the Technology Development Foundation.

This action by the Turkish government in abiding by the mandates of WTO and the European Customs Union has ended the old system of investment and export subsidies. Now the balance between technology transfer and domestic technology production is expected to improve and the demand for high technology to increase, although Turkey needs to remain alert against any future barriers to international technology transfer.

References

[1] Oldham, G., Freeman, C. and Türkcan, E. (1967), 'Trends and problems in world trade and problems', Second UNCTAD Conference, New Delhi, TD/28/Supp.

[2] ES-DA Consultancy (1995), *Survey on the Technological Innovation in the Turkish Manufacturing Industry I,* TTGV Technology Development Foundation of Turkey, Ankara, February. For the survey, questionnaires were sent to 8,375 firms employing over ten workers, of which 1,297 (16.6% of the total) responded. 23% were small firms with 10–24 employees, 37% were of medium size with 25–99 employees and 40 per cent were large firms with over 100 employees.

[3] OECD (1992), *OECD Proposed Guidelines for Collecting and Interpreting Technologies Innovation Data, Oslo Manual, OECD/6D (92)26.*

DIFFUSION OF NEW TECHNOLOGIES THROUGH APPROPRIATE EDUCATION AND TRAINING

OKYAY KAYNAK* and ASIF SABANOVIC**
Bogaziçi University, Bebek, 80815 Istanbul, Turkey
***TÜBITAK Marmara Research Center, Gebze, 41470 Kocaeli, Turkey*

1. Introduction

The 20th century has been characterised by the great strides that have been made in science and technology, dominated, in the first half, by developments in hardware. During this period, the mechanical complexity, precision and speed of production machinery in industrial installations has increased considerably. This period can therefore be dubbed the 'hardware age'. Similarly, the second half of the century may be referred to as the 'software age', since the dominant power during this period was software. At the start, electronics—until then confined to communications—penetrated industrial installations and the level of automation increased considerably in the form of 'hard automation'. The development of microprocessors in the 1970s provided a new impetus to industrial evolution. It became possible and economically feasible to realise many complex functions in a simple manner, by the use of appropriate software. At later stages, a synergetic fusion of different technologies, especially of mechanical and electronics technologies and computer science, took place and, as a result, hard automation gave way to flexible automation, culminating in computer-controlled processes, which possess a certain degree of intelligence and autonomy. Thus the age of mechatronics was born. Factories incorporating such technologies are no longer 'factories of the future' but 'factories with a future'.

As a consequence of these developments, an increasing challenge exists for the developing countries for the acquisition and diffusion of new technologies. The difficulties involved are even harder when the goal is the development of their own technology. This chapter addresses itself mainly to these issues. Firstly, in order to illustrate how new technologies are developed in industrialised countries, the present day approach of high technology companies to research and development is discussed. This is followed by a consideration of the measures that need to be taken in a developing country for the transfer and diffusion of new technologies and the promotion of technological advancement, with particular reference to the role of the academic sector in this respect. As an example, the activities of Bogaziçi University are considered and experiences of the past fifteen years are discussed. The chapter concludes with the introduction of a 'UNESCO Chair on Mechatronics', recently established at this university, and its objectives in relation to the technological development of the region.

J. Kirkland (ed.), Barriers to International Technology Transfer, 99–107.

2. Technology, Fusion and R&D

The present mechatronics age is characterised by the fusion of different technologies and the short life-cycle of technological products. A number of definitions for mechatronics have been proposed in the literature [4], differing in the particular characteristics that the definition is meant to emphasise. The most commonly used emphasises synergy: mechatronics is the synergetic integration of mechanical engineering with electronics and intelligent computer control in the design and manufacture of products and processes. The characteristics of mechatronics technology are discussed in more detail in later sections.

Industrial companies in the mechatronics age have to strive hard to keep abreast with the rapid pace of technological innovation. Missing a generation of technology can mean losing competitiveness, that is, less market share, resulting eventually in going out of business. The importance of R&D in this respect is well appreciated in industrialised countries and high technology manufacturing companies are now investing a greater amount of resources than before in R&D. In the case of some companies (especially the Japanese ones), this can be more than just the capital investment. However, the effectiveness of R&D in the development of a new technology depends more on how it is defined than on how much is actually spent on it.

The traditional approach to R&D is to develop a new technology to replace an older one. This is the 'breakthrough' approach. Another, and more contemporary approach, is the fusion of different technologies to generate hybrids. In today's competitive environment, companies cannot rely only on breakthrough research, but should include technology fusion in their R&D strategy. Japanese companies are staunch followers of this principle. For example, over the last two decades, companies such as NEC, Fanuc and Sharp have developed their own well-defined technology fusion strategies. NEC is investing heavily in optoelectronics R&D, Fanuc into mechatronics and Sharp into the fusion of crystal, electronics and optics technology. In their approaches to technology fusion, three basic principles can be identified [5]:

1. The R&D agenda is market driven and not *vice versa*. Companies therefore keep a watchful eye on the market and try to convert a vague set-up of wants into a well-defined product. This is called demand articulation.
2. Companies keep a watchful eye on technology developments both inside and outside their industry (information gathering).
3. For successful technology fusion, R&D ties are established and kept with other companies in different industrial sectors.

These principles are most apparent in the R&D strategy of Japanese companies. For example, for demand articulation Sharp has established a number of Life Soft Centres in different countries, with the express purpose of collecting information on the changes in the lifestyles of consumers and ensuring that their requirements are systematically turned into product functions. In such a centre in Japan more than 50 people are employed. Brother has established a Life Research Centre for the same purpose [3]. In the case of across–industry research, statistics show that the average number of industries per project is increasing.

3. Promoting the Diffusion of High Technology

In the section above, it was asserted that the new technologies of the mechatronics age are generally of a hybrid and complex nature. It is therefore more difficult to transfer and diffuse these new technologies than before. The existing technological level of the country plays a crucial role in determining the approach to be followed and the possible degree of success. A rough scale of three different levels may be set out [2].

1. The minimum level is characterised by an awareness of technical developments, the existence of technology monitoring agents, the recognition of appropriate and relevant technologies for the country and the ability to assess, select, negotiate and utilise emerging technology.
2. The medium level is distinguished, in addition to the above, by the ability to adapt, modify and generate its own technology.
3. The high level requires the competence to articulate demand, transfer it into a well-defined product, select the appropriate technology, design and manufacture the product and commercialise it in international markets.

Most developing countries are at the minimum level and require some stimuli for the transfer and diffusion of new technologies. These can be in the form of governmental policies in the areas of:

1. publicly financed research and development,
2. incentives (in the form of tax reliefs, and so on) to large manufacturing companies to establish their R&D departments,
3. establishment of national research and technology development programmes,
4. promotion of research activities with other countries for horizontal (south-to-south) technology transfer,
5. adoption, modification and development of new technologies/products by small and medium-sized enterprises,
6. appropriate education and training.

If we consider Turkey as a case in point, most of the above measures either have been or are being experimented with, mostly under the management of TÜBITAK (Turkish Scientific and Technological Research Council and Ministry of Industry). The programmes related to items 2, 3 4 and 5 are relatively new. Another new programme involving $40 million has recently been initiated by the Turkish Technology Development Foundation through which projects are funded on a competitive basis (with a repayment in the future) for the development of new technologies, especially in small and medium-sized enterprises. These measures have helped Turkey to move well onto the medium level described above. Some industrial sectors, especially those which manufacture white and brown goods, have now passed the manufacturing by licensing stage and are capable of developing their own technologies through adaptation and modification or original technology development research.

4. The Role of Education and Training in the Diffusion of Technology

Technological developments and education and training are interrelated in a push–pull manner. Education and training have an important role in the diffusion of new technologies and, conversely, new technologies influence education and research activities. The latter role comes about almost inherently due to the highly adaptive and receptive nature of an academic environment. However, to assist the diffusion of technology through appropriate education and training some special efforts have to be made. This is because conventional school curricula are inclined to cater more for science than technology.

The academic sector of a country can assist the diffusion of new technologies in that country in the following ways:

1. by designing curricula in such a way that the graduates are appointed with an appropriate technological profile,
2. by providing continuing education courses for practising engineers,
3. by formulating joint research projects and assigning thesis topics according to the requirements of industry,
4. by establishing technology transfer centres, such as incubation centres, innovation centres and technological parks.

Among the items listed above, the last two are more advanced levels of university–industry cooperation and can effectively be implemented only when the industry reaches a certain technological level. In other words, the demand should originate from the industry, that is, it should feel a necessity for the development of its own technology. The first experiences in this respect have a special importance. A failure of a university in meeting the expectations of industry can result in a serious setback in the further development of activities for technological transfer.

In the rest of this chapter, the activities related to the first two items of the above measures are stressed and the experiences of Bogaziçi University in this respect are described. The discussion starts with a deliberation on the characteristics of mechatronics technology and continues with the breadth versus depth profile that it requires from the technical personnel for the diffusion of the technologies involved.

5. Fusion of Different Technologies in Mechatronics

Towards the end of the 1970s, the term 'mechatronics' was given by the Japanese company Yaskawa to the integration of mechanics and electronics. However, the word has taken on a wider meaning since then. It is now understood to be an interdisciplinary engineering area in which many technologies are fused and which requires special approaches and technologies in the design and manufacture of industrial processes.

A graphical representation of a mechatronic product is given in Figure 1. It shows how the three main disciplines are fused and the depth of fusion being dictated by outside factors (which can be grouped as market demands, administration and manufacturing practices and education and training). The recent tendency is to include optics in this figure as a fourth circle to emphasise the importance of this discipline in mechatronics.

The spectrum of mechatronic products is very wide, covering

1. products having a high electronics density, for example, televisions, computers, video players,
2. products having a high mechanical density such as sewing machines, cameras, textile

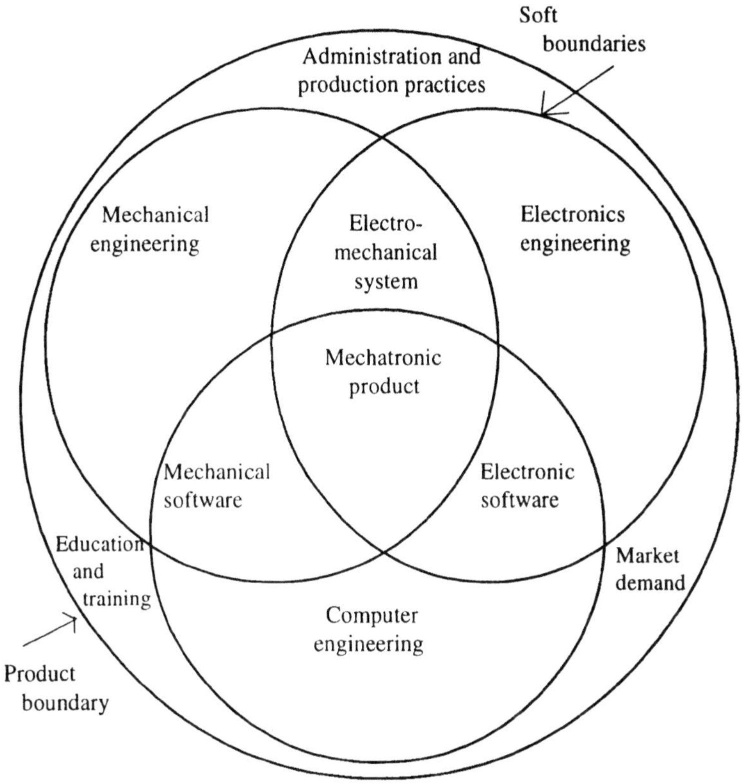

Figure 1. Schematic representation of technology fusion for mechatronics

machinery,

3. production tools such as CNC machines, robots, FMS cells,
4. components such as electrical and mechanical actuators, intelligent sensors.

It can therefore be said that mechatronics embraces almost all the industrial products of our century. This is why in the introductory section of this chapter, the phrase 'mechatronics age' was introduced.

6. Technical Breadth v. Depth Profile Essential in the Mechatronics Age

In mechatronics a number of different technologies are deeply integrated. It is therefore necessary that the technical profile of personnel involved in the design, manufacture, operation and maintenance of mechatronic products has some specific characteristics. If we look at the past, the technical personnel of the early 1900s were generalist, having a broad spectrum but not depth as shown in Figure 2. An engineer usually meant somebody that was a master of all trades, able to tackle any problem. With an increasing level of technology in industrial processes, the depth of knowledge had to increase too and this happened at the cost of breadth, resulting in distinctive specialisation areas as indicated in the same figure. However, the amount of overlap between these areas is only marginal, so that the specialists in one field can barely communicate with those of another. Such specialists may be referred to as non-communicative specialists.

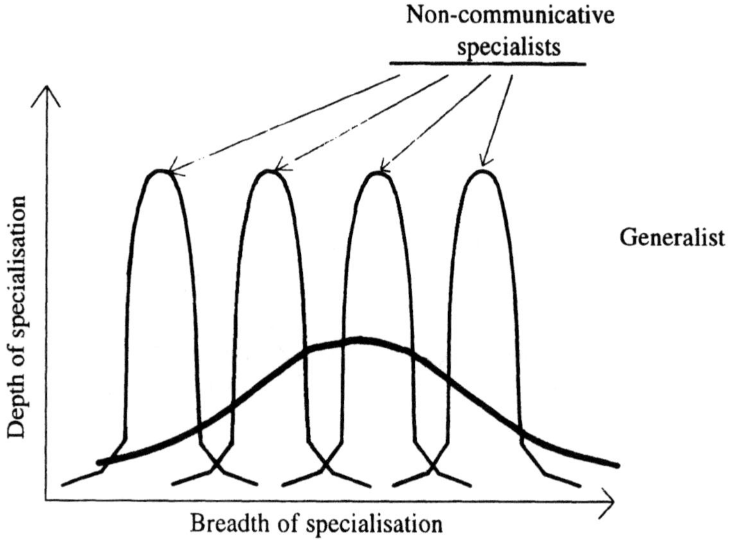

Figure 2. Depth v. breadth profile of non-communicative specialists and generalists

The technological advances of the second half of the century have necessitated an even greater depth resulting in the present sub-specialisation areas such as communication, control, power, and so on. Nevertheless, there is a considerable amount of overlap between these sub-disciplines, as shown in Figure 3, enabling communication between different specialists. However, the overlap between the main disciplines is very limited, as a consequence of which a mechanical engineer can hardly communicate with an electrical engineer. This is a serious detriment to the diffusion of mechatronics technology. The technical personnel of the mechatronics age should possess a different technical profile. The Japanese explain this in a special way, as shown in Figure 4. Technical personnel who have been educated with a classical curriculum have an 'I' type of profile, that is, they are specialists in only one area; electrical and electronics (Elect) or mechanical (Mech) engineering or computers and programming (Comp). However, a person that will design or manufacture mechatronic products has to have a 'T' type of profile, where the main stem of the T is either 'Mech' or 'Elect' and the beam balancing over it has the remaining two disciplines. Such a person is capable of communicating with people with similar profiles. A more desirable but more difficult to acquire profile is the 'π' type of profile in which the person is equally specialised in 'Elect' or 'Mech' fields but also has a 'Comp' specialisation bridge over them.

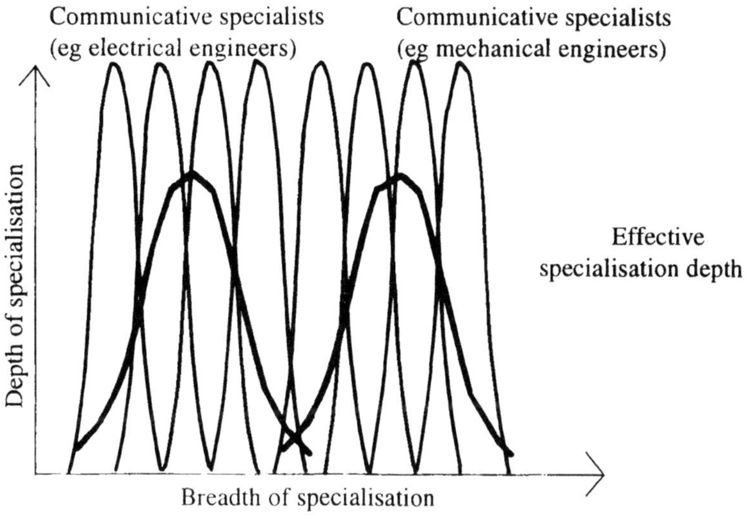

Figure 3. Depth v. breadth profile of communicative specialists

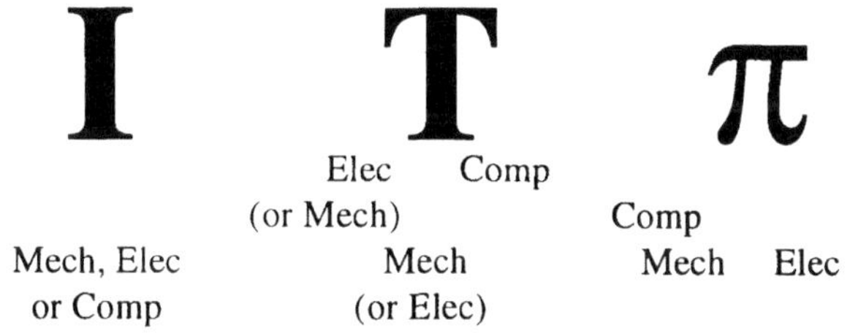

Figure 4. Specialists with different educational backgrounds

7. Activities of Bogaziçi University to Assist the Diffusion of New Technologies

Bogaziçi University is one of the leading universities of Turkey. It was founded as an American style college in 1863 (the first American college outside the USA), named Robert College. With the transfer of the site to the Turkish government in 1972, Bogaziçi University became the direct heir not only to the excellent facilities of Robert College but also to its distinguished academic traditions. The curriculum is American style and English is the medium of instruction. The faculty is young and dynamic and, through close international relations, is able to follow closely the rapid changes that are taking place in technology.

It was stated in Section 4 that the academic sector of a country can play an important role in the diffusion of new technologies in that country and different approaches available in this respect were discussed. In a similar way, the Bogaziçi Engineering School has, especially over the past decade or so, earnestly striven to assist the diffusion of new technologies in Turkey. The continuing education courses that have been offered to the practising engineer have been especially beneficial in this respect. For example, the courses offered at the beginning of the 1980s on the application of microprocessors in industrial process control were received very well and had to be repeated a number of times. These were followed by similar courses on programmable controllers. Both series of courses were designed to be of a highly applied nature and provided hands-on experience with some design problems. The demand for the courses died out within a few years, indicating a complete diffusion and absorption of microprocessor technology by industry.

Another example of the measures taken by Bogaziçi University to support the diffusion of new technologies is the establishment of a 'Graduate Program in Systems and Control Engineering'. For the time being, it offers MSc degrees only and is open to engineering and science graduates who would like to get a masters degree in the interdisciplinary area of Systems Analysis and Control Theory. Over the past seven years, the programme has had students with first degrees in Electrical and Electronics Engineering, Mechanical Engineering, Computer Engineering, Aeronautical Engineering, Industrial Engineering, Chemical Engineering and Chemistry. The graduates are well received by industry and the feedback is such that it supports the original idea behind the establishment of the programme.

The latest endeavour by Bogaziçi University to help the technological advancement of the country is the establishment of a UNESCO Chair in Mechatronics. The main objective of this chair is to promote an integrated system of research, training, information and documentation activities in the field of mechatronics and thus ensure a rapid diffusion of mechatronics technology. With this objective in mind, a mechatronics laboratory is planned to support teaching and research activities, as well as being a technology demonstrator for the industry. It is proposed that a set of enabling technologies should be incorporated, leading to the development of a sequence of research demonstrators which collectively display the functional attributes of advanced mechatronic systems. These research demonstrators will subsequently be the building blocks of industrial technology demonstrators, applicable to specific application areas important for the region. Textile, leather, small batch manufacturing and shipbuilding are among the target industries.

The UNESCO Chair is intended to pay particular attention to developing close relations with the interested parties of the region (mainly those from central and eastern Europe, the Black Sea region and the Middle East) by providing their technical personnel with continuing education courses, training in the use of high technology automation equipment, developing joint research projects, and so on. The mechatronics laboratory will be open to visitors as a technology demonstration centre. It is hoped that all these activities will be instrumental in the diffusion of high technologies in the region and thus raise the technological level of manufacturing industries.

8. Conclusions

Following the hardware and the software ages of the 20th century, we are about to enter into the mechatronics age of the 21st century, the main characteristic of which is the high level of integrated technologies that it contains. It is therefore more difficult in this age to transfer and diffuse new technologies and some special provisions have to be made. In this chapter, it is asserted that appropriately educated and trained manpower is a prerequisite in this respect. The academic sector of a country, therefore, has a special responsibility to ensure that the curricula and research and training programmes aim to achieve a certain depth versus breadth profile befitting the requirements of the century. Ability to correctly monitor, assess, select and utilise new technologies may be held to be a minimal requirement.

Another supportive measure that the academic sector can take for the diffusion of new technologies is the establishment of technology demonstration centres. These can incorporate emerging technologies and display the functional attributes of advanced mechatronic systems to industry. Additionally, they can act as a platform on which joint research and development activities are undertaken. It is hoped that the activities of the UNESCO Chair of Bogaziçi University will fulfil these aims.

Acknowledgement

The first named author acknowledges the support of UNESCO.

References

[1] Bradley, D.A., Dawson, D., Burd, N.C. and Loader, A.J. (1991), *Mechatronics*, Chapman and Hall.

[2] Bromley, A. (1990), 'Most promising applications of new technologies in industry', in F.A. Deghestani and S.A. Qasem (eds), *New Technologies and Development of the Muslim World*, Islamic Academy of Sciences, Amman, pp. 55–69.

[3] Buur, J. (1989), *Mechatronics Design in Japan. A Study of Japanese Design Methods and Working Practice in Japanese Companies*, Institute for Engineering Design, Technical University of Denmark, IK Publication.

[4] Hewit, J.R. (1992), 'Mechatronics', in M. Acar (ed.), *Mechatronic Design in Textile Engineering*, Kluwer Academic Publishers, Dorrecht, pp. 1–26.

[5] Kodama, F. (1992), 'Technology fusion and the new R&D', *Harvard Business Review*, July–August, pp. 70–78.

ECONOMIC CONSTRAINTS ON TECHNOLOGY TRANSFER TO THE TRANSITIONAL ECONOMIES:

The example of the agricultural and food sectors

A.S. DAVIES*, L.C. CRONBERG** and W.A. KERR***[1]
Economist, Agricultural & Rural Economics Department, Scottish Agricultural College, Aberdeen, Scotland.
***Managing Director, Agro-Processing Development, Malmö, Sweden.*
****Professor of Agricultural Economics, Department of Economics, The University of Calgary, Calgary, Canada.*

1. Introduction

Examination of the capital equipment of farms and food processing companies in Central and Eastern Europe (CEE) and the former Soviet Union (FSU) reveals it to be, for the most part, outdated, dilapidated, energy inefficient, unhygienic and incapable of producing high quality output. The output produced with such equipment contrasts sharply with the high quality, well packaged imported products that fill retailers' shelves. It is, therefore, not surprising that Ministry of Agriculture officials and managers of farms and food processing companies in these countries indicate that their biggest problem is the poor technology with which they have to work. Such statements must, however, be contrasted with the fact that, for example, within the Soviet Union investment in agriculture had reached 27 per cent of total investment in 1982 [13, 39]. The implication is that there were constraints on the correct planning of food system investment and subsequent use of the capital equipment provided.

It is argued in this chapter that many of these constraints have continued to exist during the process of transition from central planning to the market economy. Unless these constraints are systematically removed concurrently with any investment in new technology, the mistakes of the past are likely to be repeated. Distortions which continue to exist in the economies will result in some investments which are potentially viable either not being undertaken or unable to generate sufficient revenue to pay for themselves. Likewise, other investments which are not economically justifiable will be made. In the latter case, either the companies will be forced into insolvency or taxpayers will be forced to provide subsidies. Hence, simply making better technology available through transfer programmes will not lead to a modern efficient food system.

Of course the countries of CEE and the FSU are extremely heterogeneous. Not all the constraints outlined below apply to the same degree in all the countries of these regions.

J. Kirkland (ed.), Barriers to International Technology Transfer, 109–132.

More progress has been made in countries such as Poland, Hungary and the Czech Republic, which are at the vanguard of reform while little has yet changed in the countries of the FSU.

2. Technology Problems in the Agricultural and Food Sectors

Domestic manufacturers of farm and food processing equipment in CEE and the FSU were required to produce the number of machines set out in the formal plan for the industry. Poor design, low quality steel, and shoddy workmanship were endemic because plants were evaluated solely on the basis of whether or not the pre-established quota was achieved. The subsequent performance of the equipment has been correspondingly low: the fuel consumption of vehicles is high, and machinery productivity low. Both continually need repair. Obtaining spare parts from manufacturers has always been a problem, so each farm or processing plant had large repair shops which even manufactured equipment themselves if they were gifted with some interested worker or engineer.

In an attempt to ensure that the development of technology kept pace with that in the West, governments also imported technology. The evaluation of the new technology to be installed was usually undertaken by senior professors in research institutes and universities. However, since there were restrictions on travel to the West, even for senior technicians, most of the information concerning available technologies came from exhibitions and seminars held in CEE and the FSU by the main western equipment suppliers. Based on budget quotations from suppliers, allocations of funds were made in the five-year plans and during the five-year period further evaluations of suppliers were made through selected visits to the West.

This system had several deficiencies for the optimal selection of technology. The research institutes undertaking the selection had little practical experience. The contractors offering and quoting equipment were mainly the larger equipment suppliers who were able to afford the lengthy process of participating in exhibitions, waiting for the allocation of funds in the five-year plan, receiving different delegations and having numerous discussions in CEE capitals and/or Moscow. In order to cover the associated high transactions costs, the technology offered was normally not overly advanced and price levels were high. Investment decisions were often not made with the objective of improving the competitiveness of an industry, but rather were based on relations between suppliers and decision makers.

The low quality of raw materials, especially steel—used in the construction of farm and food processing equipment—the variable workmanship, and the need to repair it at farm or enterprise level—often without spare parts or qualified mechanics—resulted in much of the equipment being large, heavy and inadequately engineered so that parts fitted together poorly. With farm equipment this has caused soil compaction, mobility problems in inclement weather conditions and low yields. With food processing equipment it has resulted in low quality products. The poorly fitting parts cause food products to become trapped in the machinery. This, together with cleaning difficulties arising from both poor design and a lack of control, compromises hygiene.

For instance, the lack of sanitary milking systems, together with cooling and collection tanks which are difficult to drain and clean, results in low quality milk. Low quality milk makes low quality butter, cheese and milk powder. These problems are further compounded in the milk processing plants. Although a number of large milk processing plants were imported from the West and subsequently copied, the quality of domestic reproductions did not match that of imported equipment. The operation of centrifugal milk separators is adversely affected by the poorly fitting parts. The cleaning of heat exchangers is difficult due to the poor quality of rubber seals. Similarly, relaxed hygienic supervision of slaughterhouses results in low quality meat products. The shelf life of bread is also considerably less than that in the West due to the poor hygiene standards in its production, distribution and handling and insufficient levels of anti-mould chemicals.

Much of the equipment used in the food processing industry was designed when energy costs were perceived to be low. Consequently little effort was made to design energy efficient plants. This problem is compounded by the dual purpose boilers used in food processing plants. They are used to partly provide the plants with steam and partly to provide surrounding apartment buildings, normally used by factory workers, with heat and hot water. A low priority is placed upon the reduction of energy consumption when a major part of the steam produced goes to factory employees who are unable or unwilling to pay the actual cost of the energy used.

Managers in CEE and the FSU have traditionally desired sophisticated equipment using up to date technology. In part, this was in order to reduce the effort required and the benefits obtained by the workforce since the workforce selected the managers.[2] Of course it was also the desire, which is seen the world over, for the status symbol that new technology represents. However, even where the latest technology was installed in CEE and the FSU it was not used to good effect. Several years ago, a modern, computer controlled dairy farm was built outside Moscow using advanced imported technology. The feed rations to each cow were supposed to be measured and controlled by computer based on that cow's actual and expected milk production. However, despite considerable training, the system never worked as designed and was generally operated manually. Similar problems have been encountered with wet feeding systems on large pig breeding farms. The valves controlling the automatic feeding system would often break down. The manual control of the automatic system would increase the losses of feed, and in some cases caused more feed to be supplied to the pigs than they required. The consequent residues in the feed bins spoiled very easily and created health hazards.

3. Constraints on Private Investment in New Technology

As a result of these problems the technology utilised in the agricultural and food sectors of CEE and the FSU significantly lags that used in the West. The rise in input prices and the low quality of produce means that the output of many firms cannot compete with imports. Unsurprisingly, therefore, farmers and the managers of food processing companies desire new technology which they believe will allow their firms to become competitive. However, much of the new technology that they wish to acquire is embodied within capital inputs. The adoption of such technology has, of course, long-term implications.

During the process of transition from command to market economies, however, the underdevelopment of the market economies, and the incentives of the owners (and managers) of firms, can lead them to undertake investment in new technology which is inappropriate and unsustainable in the long-term.

The term 'appropriate technology' can encompass a wide range of ideas [2]. Here 'appropriate and sustainable *investment*' is used to denote investment in new technology which is capable of producing sufficient desired output in the future, both marketed and, as appropriate, non-marketed, to adequately reward society for foregoing current consumption. When the prices of inputs and outputs reflect the values placed upon them by society, this condition will be fulfilled in those instances where investment projects are capable of generating sufficient revenue so that the investor can (or, where own funds are invested, could) repay the cost of the investment at the market rate of interest. In such circumstances, the market rate of interest is taken to indicate the time preference of consumption. When there are distortions in input and/or output prices so that they do not reflect society's valuation of the resources, then the discounted social valuation of the benefits produced from the investment should be greater than the discounted social cost of the investment after appropriate adjustments for risk.[3]

There are a variety of reasons why owners and managers of farms and food processing companies may invest (or wish to invest) in new technology which is inappropriate and unsustainable. First, during transition, investors may face or perceive that they face different incentives from those that prevail in market economies, that is, the requirement to satisfy the wants and needs of consumers. These distortions can be compounded by a lack of experience with the workings of market economies as well as the industrial structure of the farm and food processing sectors inherited from the former command systems. Second, due to the underdevelopment of the institutions required for the efficient working of a market, internal prices may not reflect the true valuations of inputs or outputs. Public policy and the problems associated with the transition of the public sector can also distort private investment opportunities. Lastly, the macroeconomic environment, together with the underprovision of credit, can inhibit private investment even though it would be potentially profitable.

3.1 ENTREPRENEURIAL AND MANAGERIAL INCENTIVES

When questioned, private entrepreneurs, including farmers, in CEE and FSU countries tend to express a profit motivation. However, the incentives they face in the short-term could result in the adoption of technology which in the long run jeopardises the achievement of this objective.

3.1.1 *Moral Hazard*

Since much of the new technology desired by the entrepreneurs and managers in CEE and the FSU requires capital investment, its acquisition must be funded by increased owner equity and/or debt. Due to a lack of funds available for equity investment, the majority will, undoubtedly, be financed through borrowing.

Moral hazard arises, however, in the debt financing of investment because lenders have imperfect information about the actions taken by and, hence, the repayment ability of borrowers [23]. Consequently, in developed market economies, borrowers are encouraged to repay by loan contracts that punish default with foreclosure on collateral [4] and by legislation which, in the event of default, allows the liquidation of a firm's assets to cover its debts. These financial sanctions are designed to encourage entrepreneurs to self-select out bad risks.

However, although banks in CEE and FSU countries require significant collateral on loans and the governments of these countries have generally adopted insolvency legislation, these sanctions often do not produce this result in the transition economies. This is particularly true in the case of managers in enterprises established in the communist era whether or not they are still state owned. There are several reasons for this. First, the valuation of the assets of firms in the transition economies is very difficult. Thus, while the nominal value of a firm's assets may exceed the value of the loan, the actual value of the collateral is often less than the loan because its salvage value is low since no market exists for the assets. Second, the institutional framework required to register liens on assets tends to be underdeveloped, particularly in the FSU. It is therefore possible that the same assets can be pledged as collateral for several loans. Third, although insolvency legislation has been passed, in practice it is difficult for private creditors to institute insolvency procedures against state owned or former state owned enterprises. This is because the private legal framework to implement these procedures has failed to develop. Consequently, the responsibility for liquidating firms usually still rests with sectoral ministries. The sectoral ministries, however, have no incentives to liquidate firms. Indeed, there can be positive disincentives for ministry staff to initiate liquidations as the development of a decentralised market economy decreases their political authority and power. Further, liquidation would mean the removal of the enterprise managers with whom they had close working relationships under the command economy. Political pressure is also often applied to forestall liquidation to prevent the political and social unrest which can result from a wave of insolvencies. In the case of agro-processing firms and former state or collective farms, such disincentives are reinforced by the understandable desire of ministries of agriculture to protect the integrity of the food chain for reasons of food security.

The net result is that creditors have little ability to influence managerial behaviour [33]. Hence, when adopting new technology, entrepreneurs and managers need not give sufficient attention to the ability to repay the debt which finances its purchase.

3.1.2 *Principal-Agent Problem*

In addition, in situations where the management rights of owners are devolved to professional managers, those managers may have little incentive to maximise the return to owners. Thus they are likely to continue to select new technology which achieves their personal goals of reducing their effort and obtaining status symbols.

This principal-agent problem is particularly prevalent in situations where privatisation has resulted in fragmented ownership. One way of overcoming the problem is to utilise remuneration schemes for managerial staff that provide incentives for them to pursue courses of action which further the objectives of owners [28]. However, Tchetverina [36] notes that a result of the 'levelling ideology' of the 'real socialism' practised by the Stalinist economic system was that the economic incentives to work were repressed. Economic incentives were replaced by the evaluation of managerial performance on the basis of the fulfilment of planning targets.[4] Although later reforms[5] instituted incentive systems, as these very much related to planning targets they were still inappropriate for a private incentive based market economy. Managerial remuneration in the privatised former state enterprises, therefore, remains largely unrelated to company performance[6] (a problem which is compounded when accounts are not published).

3.1.3 *Privatisation and Ownership*

Given that compulsion was the main motivation for labour under the command systems of CEE and the FSU [36], it is probably inevitable that during the privatisation process that swept the CEE and FSU countries, workers and managers of state owned enterprises demanded large stakes in the ownership of their companies [25]. The resulting insider ownership is an obstacle to the restructuring of former state enterprises. Employee-owners resist corporate restructuring that involves redundancies, particularly when unemployment is increasing. Shedding labour is particularly difficult in industries such as agriculture which are prone to long-term decline and where the employees lack transferable skills. Moreover, the low standards of living under the communist regimes, the current abundance of previously unobtainable imported consumer goods and the uncertainty of the future causes manager and employee owners to raise wages even when it diverts resources from investment. Available funds have been diverted to current consumption, thus jeopardising the future viability of privatised enterprises.

On the other hand, insider owners desire high technology capital investment in order to reduce their own work effort. The inevitable trade-off between capital and labour is unacknowledged. In many cases former state enterprises are overstaffed even with their antiquated technology. Further reductions in staffing will be required if new technology is introduced. There is also little recognition that, in the presence of low wages, investment in costly high technology equipment makes little economic sense [27]. Insider ownership, therefore, results in the desire for effort-reducing new technology yet raises the production costs and reduces the funds available for productive investment.

3.2 LACK OF COMPETITION IN THE FOOD CHAIN

Entrepreneurs' and managers' financial motivations are, however, becoming more important as the process of transition continues. Given such motivation, the prices of inputs and outputs are crucial determinants of their production decisions. Economics suggests that, in the absence of external production effects, when there are a large number of buyers and sellers who have good information on the quantities and prices of goods being traded, and there are not barriers to the production of certain commodities, the prices of inputs and outputs will be determined by the competing desires of consumers. Thus consumers' desires determine production.

In the transitional economies these conditions do not apply and domestic prices are often radically different from those on international markets. This is particularly the case for basic agricultural products, the prices of which are often below world market prices. Although in part these lower prices stem from lower quality, a large part is the result of a lack of competition in the food chain.

3.2.1 *The Structure of the Food Processing Sector*

To fully comprehend the constraints on successful private investment in new technology in the agriculture and food sectors of economies in transition, it is first necessary to understand the organisation of industry in the former 'command system'. It is useful to think of the agro-food system in a command economy as parallel chains running from agro-inputs through primary production, processing and distribution to final purchases by consumers [17]. Each 'command firm' (in reality only a management structure designated by the planning ministry to undertake a finite number of operations) which comprised the chain was 'tied' to both its suppliers and customers. Managers were not allowed to seek out alternative suppliers when the required quantities or qualities of inputs were not forthcoming and were not allowed to locate alternative customers when surpluses were available. The planning ministry organised transfers between command firms in the chain. Given this structure, there was no need for markets or market information systems and managers gained no experience with organising transactions. The problems created by this organisational structure for the process of transition are compounded by the 'bigger is better' philosophy which was one of the main tenets of Soviet economic thought. Investments were made on a large scale so that agro-input industries, distribution systems and retail outlets tend to be geographic monopolies.

These monopolies present particular problems for the restructuring of the agricultural and food sectors. Despite low wages, the lack of demand for their products and the inefficiency of production often results in these monopolies being unable to compete effectively with imported products. However, since they are still large employers, the social and political consequences of their closure are often perceived to be unacceptable. Yet, due to their sheer size, were they re-equipped with new technology and provided with appropriate management training they would dominate local markets to the detriment of primary agricultural producers and consumers alike.

3.2.2 *The Structure of the Farm Sector*

In contrast to the slow pace of restructuring in the food processing sector, the initial restructuring of farm sectors of CEE and FSU countries has proceeded much faster. However, the approaches adopted to restructuring have been dictated by political and social pressures as much as the need for farm sectors that can efficiently produce the commodities demanded on domestic and international markets. Conflicting pressures from dispossessed landowners and their heirs for the restitution of their land, and from employees on former state and collective farms for the right to the land and farm assets, have resulted in a sharp polarisation of farm size within as well as between countries. In some cases a fragmentation of farming has resulted from the distribution of land to users and the restitution of land to former owners whilst in others, although the land and/or assets of the former state and collective farms have been privatised, the large farm units remain essentially intact—although renamed agricultural production cooperatives or something similar. Both situations present their own problems for the adoption of new technology.

In general, where the former state and collective farms[7] of CEE and the FSU have been privatised with little prior restructuring, it has been accomplished through the adoption of partnership or closed stock corporate structures (see for instance: Cook and Davies

[9] for information on Lithuania; Brooks and Lerman [7] on Russia; and Lerman, *et al.* [24] on Ukraine). These large labour owned farms still suffer from many of the problems of the original state and collective farms. They are overly staffed with rigid job demarcations and pay differentials. In addition, the productivity of labour on these farms is low because the spatial dispersion of farming results in high transactions costs to monitor hired farm labour [18]. This low labour productivity is not counterbalanced by other significant economies of size. Indeed, these large farms suffer from managerial diseconomies [12].

Although the introduction of new higher technology equipment would inevitably raise the productivity of labour, if the capital is to be utilised to its potential, restructuring of the workforce together with the introduction of job flexibility and performance enhancing incentives is required. The social consciences and lack of appropriate managerial training (at times combined with a lack of commitment to the reform process), however, makes this difficult for many farm managers to accept. Those that do accept it are confronted with another problem: the members of the workforce that require restructuring are often the farm's owners. These owner-employees combine to vehemently resist attempts to reduce their numbers and/or introduce remuneration systems that would force them to work harder.

Agricultural production cooperatives will also have difficulties funding investment in new technology [12]. On top of the practical problem, that few workers on former state or collective farms have savings available for investment in farm equipment, members of cooperatives tend to be unwilling to invest in equipment that would last beyond their own membership in the cooperative. Thus, external debt financing of investment will have to be more heavily relied upon by cooperatives than other businesses. This results in greater risk due to poor debt to equity ratios.

In general, over 20 per cent of the total labour force of the CEE and FSU countries was employed in the agricultural sector [11]. Consequently, those countries that have decided to distribute the ownership of land to farm workers now have a large number of small farms. This has also happened when land has been restituted to former owners and their heirs. Poland also has a plethora of small farms despite being the one former socialist country in Europe that retained a private farm structure.

Although these small private farms overcome the incentive and labour problems of the former state and collective farms which remain intact, they also have features which inhibit the transfer of technology. Principal among these is, of course, their small size. Like farmers the world over, those in CEE and FSU countries express a desire to have their own machinery. Although many recognise that larger equipment may have to be hired or shared amongst a cooperative, there is not yet a realisation that many of the farms are too small to generate sufficient income to justify the acquisition of even small-scale mechanised farm implements. Even where farm machinery investments are made by a cooperative association of individual private farmers it is important that the machinery is appropriate to small farms and that the payments made for its use will be able to meet its full replacement cost.

Due to their lack of savings, many of the investment needs of the small private farms are going to have to be financed by debt. Lenders will, however, demand collateral for loans. Farmers' land should be able to fulfil this role. For this to be the case, appropriate legislation is required to define property rights as well as to establish procedures for the registration of land ownership and liens. The physical mapping of farms is also required. The complexities and interrelatedness of these tasks have resulted in severe delays in the development of the framework necessary for the trade in land. Yet until land markets develop, bankers have little on which to base land valuations and are unwilling to accept farmland as collateral. In addition to these legal and technical difficulties, the development of land markets in some countries is being constrained by the unwillingness of landowners to sell. This is particularly the case where former landowners or their heirs have just regained land expropriated nearly fifty years ago. In many cases they fought legal and administrative battles to regain their property and have an emotional attachment to it.

3.2.3 *The Absence of Market Intermediaries*

In market economies a large number of market intermediaries exist to support even simple transactions. For example, a livestock farmer in the United Kingdom has a number of avenues available when he wishes to sell his finished livestock. He can take them to the local auction (a private sector market intermediary), he can sell them directly to a number of abattoirs using informal contracts, he can utilise the services of an electronic auction company, he can consign his cattle to a dealer or agent or he may even be able to make an arrangement directly with a local butcher. All of these compete for his business.

In the former command system these market institutions did not exist, because each enterprise was simply a link in a chain. State or collective farms received inputs from suppliers and delivered their output to regional processors as directed by the planning authority. Now, however, private farmers and the managers of the privatised state and collective farms supposedly have choice in where they purchase their inputs and sell their outputs. In practice though, privatisation has just created regional monopolies in input suppliers and regional monopsonies in food processors.

The development of new small food processing companies as competitors to the privatised state enterprises is hindered by the lack of market institutions. These potential entrants are unlikely to have well developed contacts with farms and are likely to be purchasing smaller quantities less suitable for direct contracting. In the absence of markets they have little information on the availability, quality and price of products. The markets in 'market economies' are signalling mechanisms. Without those signals it is much more difficult for potential entrants to judge the likelihood of success. Their absence hinders both investment in new food production technology and the demonopolisation of the food processing sector.

Thus, if new technology is to be effectively introduced into the transition economies, the institutional framework required for the market economy must first be allowed to develop. Sadly western economists advising on the transition process have largely tended to ignore the need for these institutional developments.

3.2.4 *Poorly Developed Commercial Legal Systems*

An effective legal system is one institutional requirement for a functioning market economy. Business transactions can be organised in at least three ways: (1) visual inspection—*caveat emptor*—cash; (2) one-on-one negotiations based on the development of personal trust; and (3) contracts. The former is sufficient for simple transactions; for example, consumers buying vegetables at a physical market place. The buyer visually inspects the product; if a purchase is made there is no later redress if the product's quality does not meet prior expectations and the transaction is concluded with an exchange of cash for goods. No legal system is necessary for this type of transaction. In the food industry in former command economies, physical market places have been expanding rapidly to facilitate simple transactions. However, having both buyers and sellers travel to physical market places is very inefficient—at least in terms of the opportunity cost of time.[8]

Transactions based on personal trust can be more complex and can include performance and quality criteria, redress for sub-standard products and delayed payment. This is often how transactions between small businesses are organised. Small businesses have a limited number of suppliers and a limited number of customers. A great deal of the businessman's time and effort is expended on building up personal relationships as a basis for trust. No legal system is required for these types of transactions. Unfortunately, organising these transactions is very expensive in terms of the businessman's time and hence limits a firm's growth potential.

Large firms—with many suppliers or customers—cannot rely on the 'personal trust' model because of the high cost of each transaction. They require a functioning legal system to enforce contracts. In the economies in transition, while considerable progress has been made in drawing up and enacting commercial legal systems, it is generally recognised that little progress has been made in the crucial areas of: (1) enforcement; (2) providing resources to legal institutions; and (3) educating businesses regarding the use of the commercial law system. As a result, firms cannot broaden their supplier/customer base at low cost since they have no means to ensure that they will be paid for goods or services provided, or that they can seek redress if the contract's clauses relating to quality, performance or payment are not met. Transactions based upon contracts cannot, therefore, be used efficiently, and firms are faced with the costly alternative of building relationships based on personal trust. The absence of an enforceable commercial legal system also inhibits the development of a commercial banking/credit system as financial institutions have no legal means of ensuring that loans are secure.

If one examines the process of transition in the agricultural and food systems in the former command economies, the significant areas of growth have been in small business and physical market places where a functioning commercial legal system is not necessary for organising transactions. The former large-scale enterprises of the command system remain unproductive, surviving either by subsidies or through their ability to extract monopoly rents from other former 'command firms' in their existing 'chain'. Further, there is little evidence of successful small businesses being able to make the leap to larger- scale operations—not including the Mafia which has its own 'extra-legal' methods of contract enforcement. Without a functioning commercial legal system it is unlikely that small businesses will expand to replace the existing large enterprises left over from the command era.

3.2.5 *The Dearth of Commercial Information*

The ability of existing firms to broaden the supplier/customer base of their operations and for new firms to enter an industry requires access to low cost information. Firms require information on prices, who potential suppliers/customers might be and on the reputations of any potential suppliers/customers. In modern market economies such information is available from a large number of commercial sources. Prices are published in newspapers and trade journals. In many cases price information is obtained by simply phoning around for quotes. In transitional economies, however, many firms—particularly in rural areas—do not even have a phone. Further, there can be no telephone directories or yellow pages to help firms locate potential suppliers/customers. This is particularly a problem between regions. There are no 'better business bureaux', 'chambers of commerce' and so on, where information regarding the reputation of potential suppliers/customers can be obtained. Without these transaction cost-reducing institutions to provide information, firms face very large information costs. Unless these costs are reduced firms will not know what production technology to invest in.

3.3 THE LACK OF INVESTMENT FUNDS

Competition within the food systems of CEE and FSU countries is also constrained by the lack of funds available for investment. This acts as a barrier to new companies and resources entering the industry. There are three ways in which companies in CEE and the FSU could fund their investment plans: (1) from previous retained profits or cash injections by owners; (2) by issuing equity or bonds on capital markets; and (3) by increasing bank borrowing. Each of these potential sources, however, has its own associated problems.

The high inflation and falling living standards gave the newly endowed owners of the privatised enterprises incentives to consume any retained surpluses. In some cases this problem was exacerbated by the failure of accountants to correctly adjust the value of trading stocks and depreciating capital for inflation. The effect was that enterprise owners believed that profits were far larger than they were. The distribution of this apparent profit was really capital consumption. Thus the former state-owned enterprises have little if any retained surpluses. Few private individuals have significant cash holdings to invest as the high inflation experienced in most of the CEE and FSU countries eroded the real value of savings.

Likewise, although securities markets are rapidly being developed in the CEE and FSU countries, the uncertainties surrounding both the valuation of firms and their reporting standards increases the risk of individuals investing in traded securities. Large potential rewards are required to induce individuals to take these large risks. However, such large returns are not generated by prudent long-term investment. Rather, they arise from fluctuations in the market valuation of firms which are in turn the result of the changing psychology of speculators [22]. Given that the emerging markets of CEE and the FSU have been characterised by extreme macroeconomic fluctuations, it seems that, to use Keynes' [22, p.156] analogy, they are little better than casinos and are unlikely in the short term to be the solution to firms' investment funding difficulties [see also 29].

As a result, loans remain the most important source of investment funds. The banking sectors of CEE and the FSU are, however, themselves undergoing a painful transition to the market economy. The transformation of the banking sectors in CEE and the FSU has been undertaken by the adoption of a two or, in the case of Poland [21], three-tier system with a central bank that supervises and regulates the 'commercial banks'. The 'commercial banks' are of two kinds: state-owned or recently privatised banks formed from either the regional branches of the Soviet banks or national state banks; and new private banks.

Banks were an integral part of the planning and control system of the command economy. Thus, rather than banks extending loans to enterprises on the basis of their ability to repay the credit, the banks made financial resources available to enterprises so that they could undertake the investment called for in the plan [31]. Many of the enterprises could not repay the cost of the investment since they were intrinsically unprofitable and those that could had little incentive to do so. Thus the State or recently privatised banks carry the legacy of the past on their balance sheets in the form of non-performing and often unrecoverable loans [14]. Although the new private banks do not suffer the same legacy, they tend to be undercapitalised. Insider directed lending to insiders is also an endemic problem in private banks. In addition, both the State banks and the new private banks suffer from a lack of expertise to undertake the commercial evaluation of loan proposals. The banks, as a result, are often unwilling to lend to commercial enterprises preferring to purchase government securities [15].

These problems are compounded in the farm sector. Loans are often relatively small so that the transactions costs of servicing and monitoring the loans is very high per loan. Farming is perceived to be unprofitable because output prices remain depressed while input prices tend to have risen to world market levels. Agricultural production is, of course, also subject to the vagaries of both the weather and disease. In addition, since private agricultural production is being reintroduced (apart from in Poland) after an absence of fifty or more years, bankers have little knowledge of it and find it difficult to evaluate the feasibility of proposed investments. As noted previously, farmers are constrained in their ability to provide collateral by the lack of markets for agricultural land.

The uptake of new technology is therefore constrained by a shortage of long-term commercial credit. One response of CEE and FSU governments has been to make loanable funds available directly from the state budget—usually at a cost below prevailing interest rates. Unfortunately this only exacerbates the problem. The subsidised interest rates result in excess demand for loans and the funds are therefore rationed by ministerial committees deciding which enterprises will receive the credit [32]. The poor record of past investment can give no confidence in the ability of such committees to pick the right investments now. In this system, the enterprises that have been generating the biggest losses have the most to gain by lobbying the ministerial committee for further support. The restructuring of state banks is impeded by non-market allocations of credit and the subsidised interest rates discourage private banks from lending to agriculture. The provision of subsidised credit also impedes the imposition of hard budget constraints on loss making enterprises.

3.4 MANAGERIAL ABILITY

Despite these problems many managers sincerely wish to improve their businesses and their long-term competitiveness. The rapid improvement in the quality of processed foodstuffs produced by some CEE and FSU companies is testimony to this. For the most part, however, the management of the agricultural and food system in the economies in transition—including now privatised collective and state farms—remains in the hands of the managers put in place during the command era. Small-scale entrepreneurs in the agro-inputs, food processing and distribution system as well as private farmers do exist but represent exceptions rather than the rule. For the most part they lack training in the most rudimentary of business skills.

The managers of large-scale enterprises in the agricultural and food system remain in place for two reasons. First, they have been able to use their existing power base to advantage. With better access to information, they have been able to manipulate the new system to advantage and in many cases have secured ownership or control of the existing assets. Second, there are no obvious alternatives. While the command system may have been riddled with nepotism and cronyism, it is still true that many of those who could manipulate the system to their advantage were extremely able people. The ability to manage in the command system was largely learned on the job. The management system was hierarchical with little delegation of authority. In the absence of an independent education system where management training could be obtained, there is simply no pool of potential managers to draw from outside the existing managerial cohort. Consequently, the managers of agricultural and food processing companies do not tend to have the skills required for success in a market economy.

The managerial skills required in a modern market economy centre around: (1) the ability to identify and rapidly assess potential business opportunities; (2) the planning and, thereafter, initiation of changes in the organisation which will allow those opportunities identified as potentially profitable to be acted upon; and (3) financial accountability. None of these skills were required in the command system. Identifying and assessing potential business opportunities was not required of managers as the decisions regarding what and how to produce were made by the planning authority. As a result, for example, managers have difficulty correctly assessing the appropriateness of new technologies offered by western firms.

In the command system, managers could initiate change by issuing orders. In the new system where workers have rights and where there is no communist party to quell dissent, organisational change must be accomplished through a combination of incentives and negotiations. Managers used to issuing orders are ill-prepared for these new realities.

Meeting output targets was the primary motivation of managers in the command era. To ensure that they could accomplish this central task, there was a tendency to acquire more capital than required in order to compensate for possible breakdowns which might disrupt production. Further, overmanning was common so that there was a pool of labour available if an extra effort was required at the end of the production target period. Underemployment was common. Neither of these practices fostered a concern for financial responsibility among managers.

The absence of managers with skills suited to business in a modern market economy is a major obstacle to the process of transition. When faced with making decisions relating to the acquisition of technology, these managers may focus on the ability of the technology to simply increase productive capacity rather than on whether it will be cost effective.

3.5 BUSINESS SUPPORT SERVICES

Even in developed market economies, entrepreneurs and managers rely on a wide range of specialist business support services, ranging from the frequently used accountancy, insurance and legal services to the more strategic services provided by management and marketing consultancies. The provision of such services to smaller companies is still inadequate in CEE countries and largely non-existent in the FSU.

Although the large international consultancies developed by accountancy partnerships rapidly established a presence in the CEE states and, more recently, in the FSU, they have placed their emphasis on the production of strategic analyses of the various sectors of the different countries for multilateral and bilateral donors, the restructuring and privatisation of large state owned enterprises, the provision of auditing services with particular regard to the restructuring of the banking sector, and assisting multinational corporations establish operations in the region. This is unsurprising since it is only the larger organisations that can afford the fees charged by such consultancies.

Local consultancy companies have also been set up—often by academics. These tend to concentrate on the preparation of feasibility studies. This work is important given the lack of managerial experience in this area. However, the current approach being adopted by these consultancies is very short-term. Their objective, in many cases, is only to provide complete feasibility studies which potential entrepreneurs can use to support loan applications. Entrepreneurs and managers are rarely involved in the process. Consequently, their skills are not improved[9] and important opportunities to plan business strategies realistically are lost. In cases where the feasibility of an investment is low, rather than assisting investors adjust their business plans so as to improve the project, some consultancies minimise potential risks and make overly optimistic assumptions to support loan applications. The incentives for consultancy companies to do this are increased when payment is based upon the success of a subsequent loan application. The market evaluations contained within such feasibility studies also tend to be overly optimistic and are rarely based upon direct market research. Of course, this indicates the very low skill levels in those who vet applications based upon plans by consultants. Few consultancies provide services to assist companies to improve their marketing strategies. There is also a lack of lawyers trained in commercial law and a lack of reasonably priced accountancy services. Consequently, apart from the larger state owned enterprises and foreign firms, potential investors have little access to adequate management and marketing advice in CEE and the FSU. This hampers business developments and investment.

Businesses in CEE and the FSU also need organisations which facilitate their interaction with government departments. In all economies the government is heavily involved in regulating, monitoring and taxing businesses and business activities. In modern market economies there is a well functioning system of business organisations which act as liaison between firms and the government. Although such organisations are developing in CEE and the FSU, they are not yet fulfilling all the roles businesses require of them [19]. For example, if a western government initiates a new customs procedure which results in week-long delays for goods to clear bonded warehouse facilities, the 'Customs Brokers Association' and the 'Organisation of Industrial Importers' will quickly arrange meetings with the relevant government departments to work out solutions. If major changes to the method of collecting value added taxes are contemplated by government, extensive consultations take place with business organisations such as Chambers of Commerce, Manufacturers Associations, and so on, with the objective of minimising the cost of the changes for business. While these groups may act as lobby groups for vested interests, their most important role is to provide input into the industrial standards, regulations and inspection systems put in place by governments. This reduces problems with implementing government regulations and provides a mechanism whereby potential problems can be communicated to government. Currently, businesses in CEE and the FSU see governments as sources of favours—subsidies, border protection, preferential treatment in government contracts—and their organisations expend much of their effort lobbying for 'favours'. Until business organisations become more active in the process of reducing the transactions costs imposed by government policies and regulations, firms will not be able to conduct commercial activity efficiently. The development of this aspect of business support services is as important as improving management skills themselves.

3.5.1 *Farm Extension Services*

Farmers, like other entrepreneurs, need to be able to complement their own skills with specialist business support services. However, due to the biological nature of production, in addition to the standard services required by businesses, farmers also require detailed information and advice regarding the need for action to control diseases and facilitate production. As a result, most of the CEE and FSU countries have established farm extension services which typically support the development of farmers' farm planning and credit application skills. They also provide farm management advice and information as well as supplying other services such as book keeping.

The staff of these new extension services tend to have come from offices of the local agricultural department, the management of former state or collective farms, or the agricultural research institutes or universities. Due to the prevalence of job demarcation and the lack of employment flexibility, the skills of the staff tend to be specialised. Farm advisors have few of the financial and business skills necessary to assist newly established private farmers. Their technical and specialist backgrounds tend to lead advisory staff to recommend solutions based on better production technology, sometimes with little regard to comparative costs or the wider social and household framework of private farming. Changes in farm management practice currently receive little emphasis.

Like the managers of state enterprises, farm advisors in CEE and the FSU often approach farming as the maximisation of output rather than the maximisation of profit. Advisors in these countries also have a tendency to tell farmers how they should be farming. This top down approach is based on the hierarchical management system of the former command economy. International experience, however, suggests that extension is more effective when farmers are assisted in making better choices and finding their own solutions [40].

In order to overcome these problems bilateral and multilateral donors are sponsoring a large number of projects to assist the training of farm advisory staff and the development of farm advisory organisations. These projects have typically shown high levels of success. However, since changes in individuals' philosophies are often required, the new extension organisations will require nurturing over long time periods [35]. Moreover, as these organisations develop other changes are taking place. For instance, it is difficult for both advisors and farmers to appreciate the importance of, and potential benefits from, actively marketing their output in the absence of market institutions which clearly indicate the wants and needs of consumers. However, as marketing institutions develop and the agro-processing industries become less dominated by the oligopolistic former state owned enterprises, the experiences gained from marketing training will become increasingly relevant. In the short term, however, the absence of a fully qualified extension service hampers farmers' business and investment planning.

3.5.2 *Agricultural Research*

Much agricultural technology is embodied within products which affect the performance of plants and animals or is directly embodied within plants and animals. The benefits of such technology are, however, short-lived as they become eroded by a seemingly malevolent natural environment. In order to maintain agricultural yields there is a need for on-going research.

The agricultural research institutions of CEE and FSU countries will, therefore, play a critical role in the process of transferring such farming technology. Although agricultural research results in significant international spillover effects from one region to another [38], it is still the case that new technologies need to be tested, and often adapted, before being applied in new regions. Agricultural research organisations, however, are not immune to the process of economic transition. Throughout CEE and the FSU, research organisations are under severe financial pressure. It is common for researchers not to know whether their salaries will be paid at the end of a month, never mind when they will be able to afford the equipment they require. These financial pressures are causing even the most dedicated of researchers to critically assess the rewards available in the commercial sector or in other countries. On top of these financial and personnel pressures, the transition to a market economy and private farm structure necessitates changes in the philosophy and objectives of agricultural research organisations. Previously, in the hierarchical structure of the command economies, a linear model of innovation was utilised. In this model:

> 'There is a science-practice continuum. The sequence is basic research, applied research, adaptive research, action by subject matter specialists, extension and application by farmers.' [16].

It has been suggested that such an approach to innovation is particularly inefficient [3, 10, 16]. Little use is made of the knowledge or needs of practitioners when undertaking research and feedback is not encouraged or utilised.

In order to maximise the benefits of research when agricultural production is undertaken by private farmers, it is vital that researchers respond to the needs of farmers—not just by consulting them but rather by actively involving them in the process of knowledge generation [10]. Farm extension services play an critical role in such a process by acting as a conduit for the two-way communication between farmers and researchers. There are a number of constraints to the adoption of such an approach in the CEE and the FSU.[10] First, it requires a change in the philosophies of some researchers who consider that there is nothing that farmers can teach them. Second, it requires a systematic approach to research and technology transfer by the organisations involved. The system should be characterised by shared strategic goals, synergy, decision-making by consensus and accountability to clients and policymakers [16]. However, due to their budgetary constraints, research organisations in CEE and FSU countries find it difficult to cooperate, especially when some organisations (typically the extension service) are perceived to be expanding at the expense of other organisations (typically the research institutes). The changing power relations between ministries of agriculture, research institutes, the extension service and farmers can seriously hinder the development of an effective research and extension system. Third, as mentioned previously, due to the budgetary pressures and poor remuneration of researchers, the research organisations have often lost many of the staff most able to respond to the challenges of the market system.

3.6 PUBLIC SECTOR REFORM

Economic reform in centrally planned economies necessitates public sector reform [34]. A key role of the public sector in centrally planned economies is to direct the production activities undertaken, to regulate the supplies to enterprises as well as the disposal of their output [21]. Price liberalisation and the development of a market economy implies the decentralisation of these roles. The public sectors of market economies are required to perform different functions.

Although in CEE and FSU countries it is often perceived that market economies are without laws [6], this is far from the case. Market economies require both a clear legal basis for transactions [20] and regulation so as to foster competition between producers. Regulation is also required when the production or consumption of a good effects the provision of non-marketable goods such as pollution of the natural environment. In such cases the social cost of production (or consumption) exceeds the private costs of production (or consumption). Consequently, the process of economic transition requires that some public institutions, particularly sectoral ministries, adapt to this new role. Direct public provision of goods or services can only be justified where the benefits to the general public arising from the provision of the good or service is clearly greater than the direct private benefits to those consuming the good or service. Examples include education, fundamental research, infrastructure, defence and police services.

It is difficult for the public sectors of these countries to respond to this new situation. Some institutions, such as planning commissions, lose their *raison d'être*. Others need to be reduced in size and reoriented. Some senior officials vehemently resist these changes. Pensions have been eroded by high inflation and the lack of budgetary resources so that older officials are reluctant to retire. Their job-specific human capital gained working in the central planning system and a lack of understanding of market behaviour reduces their ability to adapt to the demands of the new economic system. These problems are compounded by the meagre salaries that many officials receive which results in a lack of motivation, the loss of the more flexible, better trained, younger staff, and the need to supplement incomes.

In the absence of public sector reform, ministry officials continue to operate as they always have—directing production and investment in the enterprises that remain in state ownership. Since much of the investment is for the production of private goods (as opposed to public goods from which all in society benefit), such investment both distorts competition for the private sector and diverts resources from the provision of public goods.

3.7 AGRICULTURAL POLICY

Although CEE and FSU countries initially included the agriculture and food sectors in their programmes of economic liberalisation, more recently there has been a reversal of this trend with many countries increasing border protection and subsidies [30]. There are two major reasons for this increased intervention in the agricultural and food sectors: (1) the high degree of protection that most OECD countries give to their agricultural and food industries which is perceived to result in 'unfair' competition for CEE and FSU producers; and (2) a desire to mitigate the social and political consequences of the transition to market economies.

As a result of their wish to join the European Union (EU), Poland, Hungary and Slovakia and some of the FSU, such as Lithuania, are implementing policies which are similar[11] to the agricultural price supports of the Common Agricultural Policy (CAP) prior to the 1992 'MacSharry' reforms. They believe that aligning their policies with those of the EU will smooth the process of becoming full members of the Community. Price supports are also believed to be a means to prevent the decline in farm incomes and, hence, are a policy which will help satisfy two government goals. Whilst it is clear that policymakers wish to mitigate the downward pressure on farming incomes, it also appears to be the case that those sub-sectors which receive support are those that most require restructuring.

For instance, the breakdown in regional trade relations (particularly inter- and intra-FSU) following economic liberalisation combined with falling consumer incomes has resulted in a decline in the relative prices of milk and beef. Producers have responded by searching for new export markets. However, developing export markets is difficult because much of the dairy produce and beef[12] from CEE and the FSU is of a lower quality than that demanded in the developed markets of western Europe. Consequently, there has also been pressure for a downsizing of the industry and a large proportion of national herds have been slaughtered. Declining incomes have spurred dairy producers to lobby for price supports. The resulting government support has attenuated the pressure for the downsizing of the sector and stimulated investment. Thus, the policies encourage investment in new technology which result in the increased output of products that consumers do not desire.

A similar situation exists in the sugar sectors of some of the countries. The refining process in these countries often relied upon imported Cuban cane, subsidised locally produced beet and cheap energy imported from other regions of the FSU. Even if these refineries were equipped with new technology they would still not be able to produce sugar at world market prices [41]. Despite this, following the adoption of policies designed to protect the sugar industries of these countries, several international sugar producers have proposed joint ventures whereby the foreign partners would acquire an equity stake in the processor in exchange for the installation of new technology. Such joint ventures would, however, require ongoing subsidisation by taxpayers and/or consumers. The subsidies would represent, in part, transfers from the CEE and the FSU consumers and/or taxpayers to the foreign owners of joint ventures.

The current trend in agricultural policies in the CEE and FSU countries is, therefore, to limit the pressure for the restructuring of the agricultural sector. This should not be surprising. It is precisely the reason that most developed countries had for adopting agricultural policies [37]. One result is investment in inappropriate new technology.

3.8 THE MACROECONOMIC ENVIRONMENT

Investment in agricultural and food production within the transition economies is also constrained by the macroeconomic environment. In the command era, at state administered prices, the demand for consumer goods outstripped the available supply. The result was rationing by queuing and forced saving [26]. Upon price liberalisation, this monetary overhang resulted in high inflation—which in some of the FSU countries exceeded 1,000 per cent [30]. In order to control inflationary pressure, governments, with the support of international organisations, have instituted tight monetary policies. In some instances the control of the money supply has been combined with pegged exchange rates. These policies have resulted in high rates of interest and possibly overvalued exchange rates.

The combination of high inflation rates and high interest rates leads to short-term difficulties for investment financing even when in the long run the investment would be profitable [18]. In addition, although overvalued exchange rates artificially lower the cost in domestic currency of imported technology, they also reduce the domestic currency price of imported processed food products and raise the foreign currency prices of exports. Of course, this reduces the competitiveness of CEE and FSU agricultural producers and food processors.

4. The Need for an Integrated Approach

To summarise, whilst the technology used in CEE and FSU countries seriously lags that used in developed 'western' economies, the problems of these economies cannot be solved by technology transfer alone. Indeed, technology transfer is not even a priority. First and foremost entrepreneurs must have an economically sound set of incentives to effectively utilise new technology. A necessary condition for establishing this set of incentives is that entrepreneurs and managers both understand and bear the full financial consequences of their investment decisions. In the absence of these conditions, investments in technology will continue to be made on the basis of that which is most effort reducing rather than that which is most appropriate.

In order for consumer desires to be adequately reflected and responded to throughout the food chain, it is necessary that markets develop. Demonopolisation of the food processing sector is required, together with the development of marketing institutions, a commercial legal system and low cost information.

In addition, if entrepreneurs and managers are to respond effectively to incentives, a variety of other constraints on their investment behaviour need to be removed. Clearly, there is a need for an increased availability of funds available for investment. Despite the development of capital markets, in the medium term the majority of this burden will fall upon the banking sectors. The banking system requires assistance with its adjustment to a market economy. It is also necessary that managers be trained in the financial and marketing skills required to realistically plan new investment. This set of skills goes well beyond the technocratic approach currently being utilised in the preparation of investment feasibility studies. Further development of the business support sectors and management training is required.

The expansion of private investment in these countries will require a further withdrawal of the state from the provision of private goods. Rather, public sector activities should be focused on the provision of public goods including education and research. Ministries of agriculture should also resist the temptation to give in to lobbying and institute policies that slow the restructuring process. Border measures, price supports and subsidies will only result in the misdirection of investment funds to the long-term detriment of both investors and the country as a whole. Finally, further efforts should be made to reduce inflation so that nominal interest rates can fall. It is, however, important to recognise that the use of exchange rate policies to achieve this objective will be detrimental to the agricultural sector. An appropriate balance needs to be set.

Without the development of all the components of a market economy, technology transfer will be largely ineffective. There are a large number of examples from the experience in developing economies to suggest that simply providing technology does not mean that it will be utilised effectively—water pumps without petrol supplies, hospitals without drugs and antiseptics, tractors without spare parts, and so on. Although the levels of technical education in the transition economies are higher, and therefore using and maintaining new technology will not be as difficult as it is in developing countries, if technology transfer is not based upon the ability of the new technology to pay for itself similar mistakes will be made in the transition economies.

The need for a systems oriented approach to technology transfer is amply illustrated by some of the private sector developments taking place in CEE and the FSU. The lack of a marketing system is overcome by direct contracting or vertical integration between retailers and producers. In order to overcome liquidity problems farms are supplied with inputs (seeds, fertilisers, and so on) on a credit basis. Technical assistance is provided on cultivation, harvesting, post harvest handling, and storage. Producers are then paid a balance after deducting for the credit extended. Such developments are, however, happening on a piecemeal basis and are usually undertaken by foreign companies. They do not remove the need for the systematic alleviation of the economic constraints to technology transfer.

Unfortunately, the systemic problems identified above require a long-term, often unglamorous, commitment of resources to rectify—far more commitment than the simple direct transfer of engineering-based technology. Improvements to enforcement in the commercial legal system, the establishment of technical standards, compiling telephone directories and educating managers are activities which have great difficulty competing for resources against flashy new machines and computer gadgetry.

The incentives for the commercial acquisition of technology which can be effectively utilised can only come from a functioning market system. This is the most fundamental lesson which was learned from the experience of planned economies. The inability to choose new technology rationally or to integrate it into the broader economy was the central reason for the command economies' inability to close the gap in living standards with modern market economies. This lesson must not be forgotten in the day-to-day struggle to transform the economies of CEE and the FSU. If it is, the process of transition will be much longer and more convoluted than it need be.

Notes

1. The authors gratefully acknowledge the long discussions of the issued raised herein with colleagues in central and eastern Europe and the former Soviet Union, especially Miss J. Mortka. The opinions expressed are, however, those of the authors with whom responsibility remains for any errors or omissions.
2. An old Russian proverb states that you would not feel that you had done a good day's work if you were not able to take something home with you that day.
3. Where the products subject to price distortions are internationally traded, border prices may be used as estimates of the opportunity cost of the products. Where products are not traded or not marketed other more sophisticated techniques must be utilised to estimate the social valuation of the benefits produced and the costs incurred.
4. The planning process however depended upon information supplied by managers which, of course, provided incentives for mis-reporting so as to reduce the effort required to fulfil planning targets [5].
5. Such as the reforms of 1971 which resulted in a scheme which has been termed the 'New Soviet Incentive Model' [5].
6. Mayhew and Seabright [28] suggested that adequate incentive systems should have been set up prior to privatisation.
7. As noted by Csaki and Lerman [11], in general the treatment of state farms has differed from that of collectives and the approaches have varied between countries. These differences are abstracted from in order to highlight key principles.
8. Indeed, the large transactions costs involved with such institutions is causing their disappearance from developed market economies.
9. Indeed, this seems to be the worry of the consultancy companies, that once managers can prepare studies themselves their services will no longer be required.
10. It should of course be noted that the CEE and FSU countries are not the only ones that have difficulties in implementing true participatory research processes. This problem is one which challenges research and extension systems worldwide.
11. The extent of this similarity differs between the countries of CEE and the FSU. For instance, in those countries where the market economy is least developed, although the description of some agricultural policy mechanisms is superficially similar to those of the EU's Common Agricultural Policy, the almost complete absence of markets means that the policies are merely influencing the bargaining behaviour of bilateral monopolies.
12. Beef produced in CEE and FSU countries tends to be a by-product of dairy systems—cull cows and steers—rather than the result of specialist beef production.

References

[1] Alinska, A., Kulawik, J. and Gajewski, G. (1994), 'Poland's farm and agriprocessing banks face restructuring', *Agricultural Income and Finance*, 54, 24–35.

[2] Amsden, H. (1989), 'Appropriate technology', in Eatwell, J., Milgate, M. and Newman, P., *The New Palgrave: Economic Development*, Macmillan, London.

[3] Bell, S. (1994), 'Methods and mindsets: towards an understanding of the tyranny of methodology', *Public Administration and Development*, 14, 323–38.

[4] Benjamin, D.K. (1978), 'The use of collateral to enforce debt contracts', *Economic Inquiry*, 16, 333–59

[5] Bennett, J. (1989), *The economic theory of central planning*, Basil Blackwell, Oxford.

[6] Bromley, D.W. (1993), 'Revitalising the Russian food system: markets in theory and practice', *Choices*, Fourth Quarter.

[7] Brooks, K. and Lerman, Z. (1994), 'Land reform and restructuring in Russia', *World Bank Discussion Paper* 233, Washington D.C.

[8] Collins, P. (1993), 'Civil Service reform and retraining in transitional economies: strategic issues and options', *Public Administration and Development*, 13, 323–44.

[9] Cook, P. and Davies, A. S. (1994), *Lithuanian Farm Survey: A Report for the World Bank and the Government of Lithuania*, Scottish Agricultural College , Aberdeen.

[10] Cornwall, A., Gujit, I. and Welborn, A. (1993), 'Acknowledging process: challenges for agricultural research and extension methodology', Discussion Paper 333, Institute of Development Studies, Universirty of Sussex.

[11] Csaki, C. and Lerman, Z. (1994), 'Land reform and farm sector restructuring in the former socialist countries in Europe', *European Review of Agricultural Economics*, 21, 553–76.

[12] Deninger, K.W. (1993), 'Cooperatives and the break-up of large mechanised farms: theoretical perspectives and empirical evidence', *World Bank Discussion Paper* 218, Washington D.C.

[13] Desai, P. (1992), 'Reforming the Soviet grain economy: performance, problems and solution', *American Economic Review—Papers and Proceedings*, 82(2), 49–54.

[14] *Economist* (1993), 'Survey of eastern Europe (4): step B: banking and budgets —is economic reform irreversible? What is needed to make it so?' *The Economist Newspaper*, 13 March, 326.

[15] *Economist* (1994), 'Relieving central Europe's banks of their burden: many banks in Hungary, the Czech Republic and Poland are inefficient and plagued by bad debts. They are trying to escape from the legacy of the past', *The Economist Newspaper*, 27 August, 332.

[16] Eponou, T. (1993), 'Integrating agricultural research and technology transfer', *Public Administration and Development*, 13, 307–18.

[17] Gaisford, J.D., Hobbs, J.E. and Kerr, W.A. (1995), 'If the food doesn't come—vertical co-ordination problems in the CIS food system: some perils of privatisation', *Agribusiness: An International Journal*, 11(2), 179–86.

[18] Hill, G.P. (1981), *The Feasibility of Financing Investments Using Borrowed Money During a Period of Inflation and High Interest rates*, Department of Agricultural Economics, Wye College.

[19] Hutchins, R. K., Kerr, W.A. and Hobbs, J.E. (1995), 'Marketing education in the absence of marketing institutions: insights from teaching Polish agribusiness managers', *Journal of European Business Education*, 4(2), 1-18.

[20] Intrilligator, M. D. (1994), 'Privatisation in Russia has led to criminalisation', *Australian Economic Review*, Second Quarter, 4–14.

[21] Keren, M. (1993), 'Optimal tautness and the economics of incentives in bureaucracies', *Comparative Economic Studies*, 35(1), 85–117.

[22] Keynes, J. M. (1936), *The General Theory of Employment Interest and Money*, Macmillan, London.

[23] Leathers, H. D. and Chavas, J. P. (1986), 'Farm debt, default, and foreclosure: an economic rationale for policy action', *American Journal of Agricultural Economics*, 66(4), 828–37.

[24] Lerman, Z., Brooks, C. and Csaki, C. (1994), 'Land reform and restructuring in Ukraine', *World Bank Discussion Paper* 270, Washington D.C.

[25] Lieberman, I. W. (1995), 'Mass privatisation in central and eastern Europe and the former Soviet Union: a comparative analysis', in *Mass Privatisation: An Initial Assessment*, OECD, Paris.

[26] Lipton, D. and Sach, J. D. (1992), 'Prospects for Russia's economic reforms', *Brookings Papers on Economic Activity*, 2, 213–76.

[27] Marjit, S. (1988), 'A simple model of technology transfer', *Economics Letters*, 26, 63–67.

[28] Mayhew, K. and Seabright, P. (1992), 'Incentives and the management of enterprises in economic transition: capital markets are not enough', *Oxford Review of Economic Policy*, 8(1), 105–29.

[29] Mullineux, A. (1992), 'Privatisation and financial structure in eastern and central European countries', *National Westminster Bank Quarterly Review*, May, 12–25.

[30] OECD (1994), *Agricultural Policies, Markets and Trade in The Central and Eastern European Countries (CEECs), the New Independent States (NIS), Mongolia and China: Monitoring and Outlook 1994*, OECD, Paris.

[31] Patterson, P. L. (1992), 'Capital markets and financial issues in economies in transition', *American Journal of Agricultural Economics*, 72, 1170–3.

[32] Petit, M. and Brooks, K. (1994), 'The role of the west in the reconstruction of agriculture in eastern and central Europe and the former Soviet Union', *European Review of Agricultural Economics*, 21, 477–92.

[33] Phelps, E. S. (1992), 'Discussion of prospects for Russia's economic reforms by D. Lipton and J.D. Sachs', *Brookings Papers on Economic Activity*, 2, 213–76.

[34] Pollack, R. A. (1985), 'A transactions cost approach to families and households', *Journal of Economic Literature*, 29, 913–20.

[35] Ragland, J. (1994), *Restructuring and Modernisation of the Polish Agriculture Sector: Lessons for the Emerging Democracies*, paper to the Europe, Africa, and the Near East Agricultural Counsellor/Attaché/ATO Conference, 27 October, Warsaw, Poland.

[36] Tcetvernia, T. (1991), 'Labour incentives in alternative forms of production', in Stadding, G. (ed.), *In Search of Flexibility: The New Soviet labour Market*, International Labour Office, Geneva.

[37] Tracy, M. (1993), *Food and Agriculture in a Market Economy: An Introduction to*

Theory, Practice and Policy, Agricultural Policy Studies, Belgium.

[38] University of Minnesota (1990), *Economic Evaluation of Agricultural Research*, IR-6 Information Report No. 90–1, Department of Agricultural and Applied Economics.

[39] USDA (1983), *USSR Review of 1982 and Outlook for 1983*, US Government Printing Office, Washington, D.C.

[40] Van Den Ban, A. W. and Hawkins, H. S. (1988), *Agricultural Extension*, John Wiley & Sons, New York.

[41] World Bank (1995), *Lithuania: Agriculture and Food Sector Review*—Volume II, Washington D.C.

PROFESSIONAL AND TECHNICAL STRUCTURES AS A BARRIER TO TECHNOLOGY TRANSFER[1]

TAMÁS G. TARJÁN
Institute of Economics, Hungarian Academy of Sciences
Budapest

1. Introduction

This chapter begins with a review of research projects carried out by the author over the last five years into innovation, technology and human capital in east central Europe. These projects necessitated the revival of Jánossy's 30-year old trendline theory. A short summary of Jánossy's book [2] is therefore given.

Jánossy divided his book into two parts.[2] In the first part he formulated and empirically proved the law of economic development—his trendline theorem. He discussed the economic development of the major economies in this century, using data available up to 1965, which proved the validity of his theorem. In the second part he gave explanatory proof by means of a search for the fundamental reasoning behind the law by which he succeeded in identifying the basic building blocks of economic development.

The question now arises whether or not the history of the last 30 years bears out the validity of Jánossy's theorem. In his 1995 paper [15], the present author extended Jánossy's time-series for another 30 years and found confirmation of the theory. Here, a summary of this work is presented, together with recent results from Maddison's database, which themselves differ only slightly from that of the Mitchell–Liesner data. It is concluded that the fundamental reasons for Jánossy's law remain valid today and, at the same time, they prove to be the most important barrier to technology transfer.

1.1 BACKGROUND TO THE REVIVAL OF JÁNOSSY'S TRENDLINE THEORY

In 1990–91 a wide ranging international comparison was carried out under the leadership of George F. Ray, a senior research fellow at the London-based National Institute of Economic and Social Research, to address the size of the technological lag of former east European socialist countries [11][14]. The research project concentrated on the state of smaller CMEA countries, similar to Hungary, in 1989. Research workers from Bulgaria, from the former Czechoslovakia, and Poland and Hungary participated. When the research was begun the rapid collapse of the statistical systems of CMEA and the eastern European countries was not anticipated. Now we know that this was a last minute effort at measurement.

The research included the most important industrial technologies introduced in developed industrial countries after World War II and we investigated the latest so-called

J. Kirkland (ed.), Barriers to International Technology Transfer, 133–145.

'high-tech' production systems and products, energy production and consumption, international patenting and research and development. From this research, started by George Ray and finished in 1991, we estimated the technological lag (both in introduction and diffusion) of Hungary's industry behind that of Austria to be about twenty years. This was an average, the figure was larger in heavy industry and smaller in light industry. It is important to note, especially in the case of Hungary, that the research did not include agriculture and the food industry. In the meantime the investigation was also extended to the agriculture and food industry [13], where the lag was found to be smaller.

In 1993–5 a European Union COST (A7) research project[3] was carried out under the leadership of János Kovács, a scientific head of department of the Institute of Economics, Hungarian Academy of Sciences. As part of this project, an international conference was held in Budapest, in January 1994, and the revised material was published and edited by János Kovács [5]. In contrast to the former project, the latter laid emphasis on the human factor and pointed out the apparent paradox that, parallel with the several-decade-long lag in productivity, Hungary has a manpower and R&D base of relatively higher quality than its actual technical level.

These various research projects, dealing with the problem of innovation, technology and human capital in east central Europe, caused Jánossy's theory to be revised today, 30 years later.

1.2 JÁNOSSY'S TRENDLINE THEORY ON ECONOMIC RECONSTRUCTION PERIODS

Ferenc Jánossy was looking for a substantial carrier of economic growth and, in the mid-1960s, he developed and published a very attractive and elegant theory [2][3]. Thus, in the course of mechanism debates (preceding the 1968 Hungarian economic reforms), a lively professional debate was elicited which had positive acceptance in Hungary.

2. Formulation of the Law

2.1 DEFINITION

The Jánossy trendline is a straight line which connects the outstanding values (local maxima) of a production indicator's time series (plotted on a logarithmic scale) over a given period of time.

2.2 JÁNOSSY'S THEOREM

If the economic development of a country involved directly in a war was undisturbed before the outbreak of that war (that is, its production indicator's time series, GDP or per capita GDP, plotted on a logarithmic scale is a straight line), then with postwar reconstruction the straight line of prewar development will again be reached and followed just as if the war had not happened, independent of whether the country was winner or loser.

2.3 ILLUSTRATION OF THE THEOREM

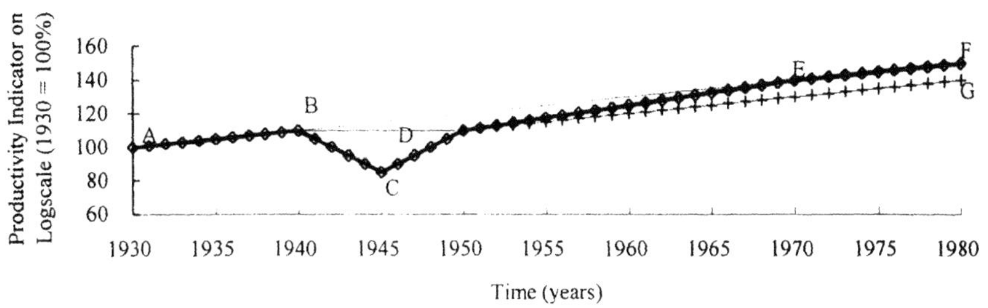

Figure 1. Postwar reconstruction period, schematic diagram

AF: the *Jánossy-trendline* of economic growth
AB: the development of production up to the outbreak of war
BC: the regression of production caused by the war
CE: the development of production during the reconstruction period namely:
CD: up to reaching the before war level
DE: up to reaching the Jánossy-trendline
EF: the real development of production after the reconstruction period
DG: the supposed development of production after the reconstruction period

With the help of Figure 1, Jánossy's theorem can be stated as follows (see [2], p.18). The reconstruction period starts at point C. From this time production is growing incessantly and after some years reaches the prewar level (point D). However the reconstruction period does not stop here because the production progresses at nearly the same rapid pace as before up to the point E, pending arrival at the Jánossy trendline of economic growth. Only now does the rate of growth halt and come back—more or less suddenly—to the normal level of growth which is typical and determined by the rules of long-run economic development.

After this point in time the production of growth follows the Jánossy trendline (see segment EF). Thus the reconstruction period will be completed much later than it is often, wrongly, supposed; that is, it does not come to an end at point D—when the reconstruction reaches the prewar level—but at point E, when the level of production will again catch up with the Jánossy trendline of economic development.

3. Explanatory Proof of the Law

3.1 EXPLANATION OF THE THEOREM[4]

Since the straight line of the production indicator (plotted on a log scale) of prewar development is followed after the reconstruction just as if the war had not happened, the process of development must imply a crucial factor which remains intact during the war (see [2] pp. 112–13). That basic building block is necessarily 'human capital'—that is, the manpower of a country with its 'professional structure'—because the other most important factor of production, the 'physical capital', has been almost totally destroyed during the war. However, in spite of the fact that manpower has been diminished too, its 'professional structure' and its stage of development has not only been preserved but has also developed without interruption. Thus the rate of growth of economic development depends after all on the development of 'human capital'.

Jánossy's 'human capital' notion may now be put into the most important 'four elements in development' (see [12] p. 885), that is:

- Human resources (labour supply, education, discipline, motivation)
- Natural resources (land, minerals, fuels, climate)
- Capital formation (machines, factories, roads)
- Technology (science, engineering, management, entrepreneurship).

'Human capital' is related directly to the first above-specified, and indirectly to the fourth element.

The relation between the change in 'professional structure' of the manpower of a country and the rate of growth of economic development is summed up by Jánossy in his following four points (see [2] p.245):

1. The level of development of a given country—even if it is only a possibility in the real magnitude of production—depends first of all on the actual professional structure of the whole manpower.
2. Economic development is linked to the change in professional structure. The preliminary condition for a faster rate of economic development is a faster change in professional structure.
3. The barriers which limit the speed of change in professional structure also limit, in the long run, the economic development rate.
4. The inertia stabilising the change in professional structure—the influence of changes in the past on the changes of the future—determines profoundly the persistence of the trendline of economic growth.

Later ([2] p.270) Jánossy gives an explicit statement concerning the relationship between new technology and professional structure, which claims that:

5. 'New technology' can influence the economic progress of the less developed countries only through their professional structures.

For us now the main message of Jánossy's work is that for long-run development the human factor is fundamental. To put it another way, between the two factors of production which mutually depend on one another it is 'human capital' (in a general sense the 'professional structure' of the manpower of a country) which is decisive. The two factors of production might have separated for certain extraordinary—shorter or longer—periods and these are the periods investigated by Jánossy.

There were other people who investigated the sources of economic growth. The most famous are the teams led by Robert Solow, John Kendrick and Edward Denison. Relying on the method of so-called growth accounting they obtained results similar to those of Jánossy. Denison investigated the sources of growth through US data for the period 1948–86:

'Studies using the techniques of growth accounting break down the growth of GNP in the private business sector into its contributing factors. These studies find that capital growth is only a modest contributor, accounting for about a quarter of total GNP growth. Education, technological change, and other sources make up almost a half of total GNP growth and six-tenths of the growth of output per worker' ([12] p.864).

4. Empirical Proof of the Law Today

4.1 THE JÁNOSSY TRENDLINE IN OECD MEMBER COUNTRIES IN THE 20TH CENTURY [15]

The question may now be asked whether the development of the world economy in the last thirty years since Jánossy's theory has been in line with or contrary to his trendline theory. To answer we make use of the OECD database which pioneered the assessment, aggregation and publication of data necessary for the quantification of the performance of the leading countries of the world economy. At present we have at our disposal 32 years of time series about the most important macroeconomic indicators of the 24 member countries [10]. We shall rely on their per capita GDPs at the price level and exchange rates of 1985 ($US) from 1960 to 1991.

4.2 CONCLUSION TO BE DRAWN FROM THE OECD DATA 1960–1991 ([10] TABLE 20.1, pp. 130–31)

After representing the time series of 24 countries on the logarithmic per capita GDP axis we can see that the actual data fluctuate—with moderate deviations—around a straight line with one point of inflexion somewhere in the early 1970s. Therefore we felt able to fit to the constant price, logarithmic per capita GDP data of the 24 OECD countries (with the standard method of mathematical statistics, the method of least squares) a broken line with one point of inflexion. The main results of the computation are summarised in Table 1.

As shown in Table 1, our broken-line estimation corroborates the visually immediate observation that the break-point is situated at the beginning of the 1970s. The trend of OECD countries taken together has its break-point in 1971 with a relative growth–recession indicator $bt_1/bt_2 = 1.99$, and that of the Common Market and of European OECD countries in 1973 with $bt_1/bt_2 = 1.97$.

In Figure 2 we see the main representative of the three geographical regions of the world and their relative development for the last three decades. In all the three cases we also depicted the fitted broken line. The break-point for the US was in 1968, for Japan in 1970, for the European OECD countries in 1973. It can be seen that the difference (measured in logarithms) between the more developed parts of Europe and the US hardly diminished during those 31 years; that is, if we take the US as 100 per cent, Europe could raise its 37 per cent level of 1960 to only 47 per cent by 1991. At the same time Japan not only overtook the European OECD countries but approached even the US; expressed in figures its development level relative to the US rose from 28 per cent to 80 per cent.

TABLE 1. The results of fitting the broken trendline with one point of inflexion for OECD data (1960–91)

Country	Break-point (year)	bt1 rate of growth (per cent)	bt2 rate of growth (per cent)	bt1/bt2 (ratio)
Canada	1976	3.69	1.94	1.90
United States	1968	3.24	1.44	2.24
Japan	1970	9.29	3.33	2.79
Australia	1972	3.51	1.57	2.23
New Zealand	1973	1.90	0.82	2.33
Austria	1976	4.22	2.02	2.08
Belgium	1974	4.27	1.89	2.26
Denmark	1969	3.79	1.95	1.95
Finland	1973	4.29	2.69	1.59
France	1973	4.34	1.77	2.46
Germany	1973	3.41	2.01	1.69
Greece	1973	7.12	1.61	4.43
Iceland	1981	4.14	1.91	2.16
Ireland	1973	3.67	2.88	1.28
Italy	1973	4.46	2.45	1.82
Luxembourg	1985	2.24	3.85	0.58
Netherlands	1973	3.92	1.22	3.22
Norway	1986	3.60	0.71	5.09
Portugal	1973	6.52	1.99	3.28
Spain	1972	5.79	1.67	3.48
Sweden	1971	3.51	1.58	2.22
Switzerland	1971	2.73	1.12	2.44
Turkey	1976	3.71	1.95	1.90
United Kingdom	1968	2.39	1.96	1.22
OECD total	1971	3.78	1.90	1.99
OECD Europe	1973	3.57	1.70	2.10
EEC	1973	3.74	1.89	1.97

Notes:
The first column of the Table 1 called break-point gives the year of inflexion
The second and third columns bt_1 and bt_2 give the rates of growth before and after the year of inflexion of the fitted broken trendline whereas in the column bt_1/bt_2 we can find the ratio of these two data.

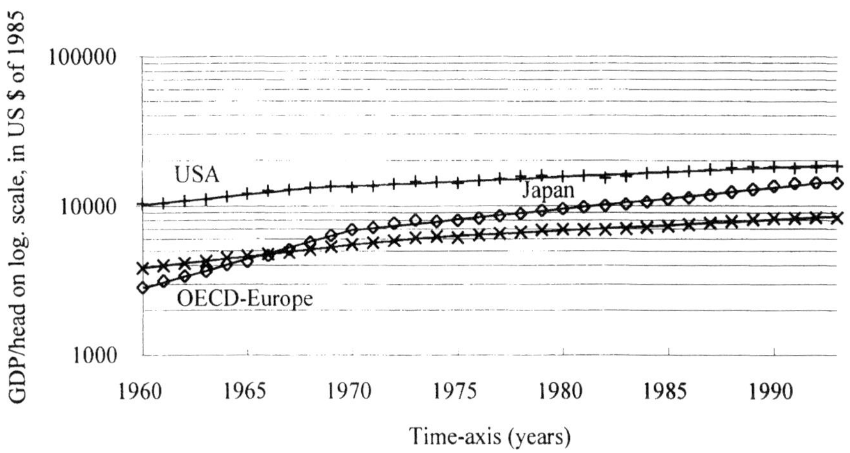

Figure 2. Development of the United States, Japan and the European OECD countries together, 1960–91

4.3 LESSONS FROM 20TH CENTURY HISTORICAL DATA

We shall now examine the existence of the development trend assumed by Jánossy in the thirteen of the 24 OECD countries for which GDP indicators are available from the beginning of this century.

We can establish that the US trendline shows a slight break in 1968, while in the other five countries (belonging to the so-called G7) which underwent a reconstruction period after the World Wars (Japan, France, the GFR, Italy and the United Kingdom) —as well as in Austria and Denmark—a straight trendline asserts itself in the 20th century. The trendline shows a break in the case of the Netherlands, Norway, Spain, Sweden and Switzerland.

Let us depict per capita GDP at 1985 constant prices ($US) for the above mentioned thirteen OECD countries on a logarithmic scale. Then let us draw in Jánossy's 20th century trendline which, by definition, connects the highest per capita GDP data. We can see the results in Figures 2–6. Table 2 summarises the empirical finding that in most cases the broken lines are parallel after the break-point with Jánossy's 20th century trendlines.

4.4 MITCHELL–LIESNER'S DATA

Let us now investigate one by one the thirteen OECD countries for which either Mitchell or Liesner have data to see whether Jánossy's theorem is valid for them and if so how accurately. (We use for this exercise Figures 2–6 and the second column of Table 2.)

The Jánossy trendline, by definition, connects with a straight line the largest (on a logarithmic scale) output data of a given period.

As we have already mentioned the trendline was conceived—before Jánossy's theory—as a kind of average, around which the data fluctuate according to a regular or irregular

TABLE 2. The comparison of parameters of the broken trendlines with one point of inflexion fitted for OECD data (1960–1991) and that of Jánossy trendlines drawn to the historical statistical data of the whole 20th century

Country	Break-point (year)	Janossy's trendline for C20th growth rate Mitchell's data (%)		After break-point growth rate bt2 (%)		Janossy's trendline for C20th growth rate Maddison's data (%)
Canada	1976	-		1.94		3.46
United States	1968	1.74		1.44		1.74
Japan	1970	3.45	=	3.33	=	3.16
Australia	1972	-		1.57		1.38
New Zealand	1973	-		0.82		-
Austria	1976	2.08	=	2.02	=	1.99
Belgium	1974	-		1.89		1.72
Denmark	1969	1.88	=	1.95	=	2.05
Finland	1973	-		2.69	=	2.66
France	1973	1.92	=	1.77		2.09
Germany	1973	2.07	=	2.01	=	2.20
Greece	1973	-		1.61		-
Iceland	1981	-		1.91		-
Ireland	1973	-		2.88		-
Italy	1973	2.28	=	2.45		2.19
Luxembourg	1985	-		3.85		-
Netherlands	1973	1.72		1.22		1.78
Norway	1986	2.67		0.71		2.64
Portugal	1973	-		1.99		-
Spain	1972	2.10		1.67		-
Sweden	1971	2.76		1.58		2.32
Switzerland	1971	1.88		1.12		2.22
Turkey	1976	-		1.95		-
United Kingdom	1968	1.88	=	1.96	=	1.88
OECD total	1971			1.90		
OECD Europe	1973			1.70		
EEC	1973			1.89		

Notes: In the first column of Table 2, called break-point, we find the year of inflexion. In the second column, we show the rate of growth of the Jánossy trendline for those thirteen countries for which either Mitchell or Liesner have data. In the third column we show rates of growth of fitted broken lines after the break-point for OECD countries for the period 1960–91. In the last column we show the rate of growth of the Jánossy trendline for those countries for which Maddison has data. We indicated with equality signs left and right from the third column countries whose rates of growth, $bt_{2.}$ in the third column differ less than 8 per cent from the rate of growth of the second and fourth columns.

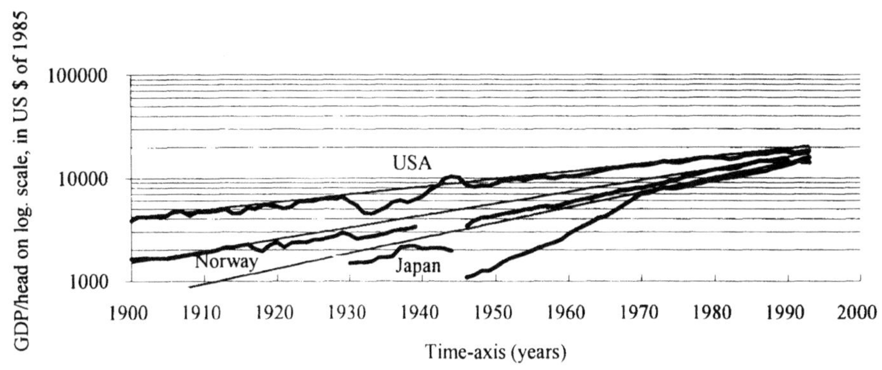

Figure 3. Development of Japan, Norway and United States in 20th Century

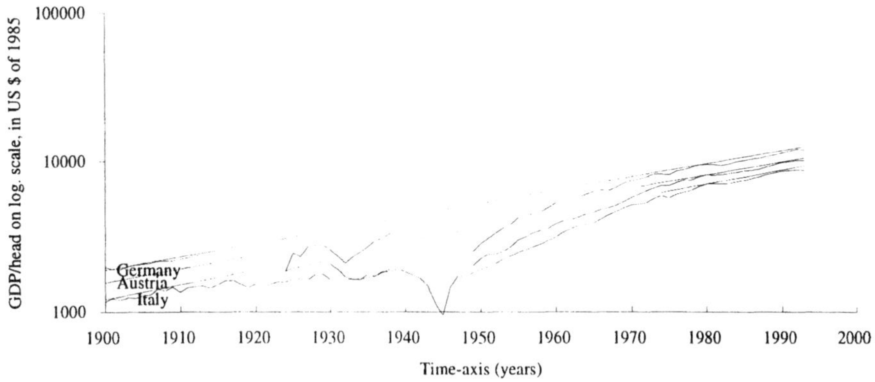

Figure 4. Development of Italy, Austria and Germany in 20th Century

cyclical movement. Jánossy's definition is different since, at the precise time of wars and subsequent reconstruction periods, a trendline defined as average would depend strongly on the extent of wartime recession, which is independent from the general production capacity of the economy. Jánossy's finding was precisely that the trendline defined through local maxima does not depend—within certain limits—on the extent and duration of the recession. This definition proved to be feasible in most cases and does not lead to inconsistencies. This is not, however, the case for the US (see Figure 2).

On per capita constant price gross output, the US has twice risen above trend during the 20th century in 1906–7 and in 1943–6. In addition, this does not take into account the rise during World War II, when the United States was in an exceptionally favourable situation relative to other countries devastated by the war.

Figure 2 gives 20th century data for the US and its Jánossy trendline. We can see from Table 2 that the rate of growth of the Jánossy trendline is 1.74 per cent, whereas the rate

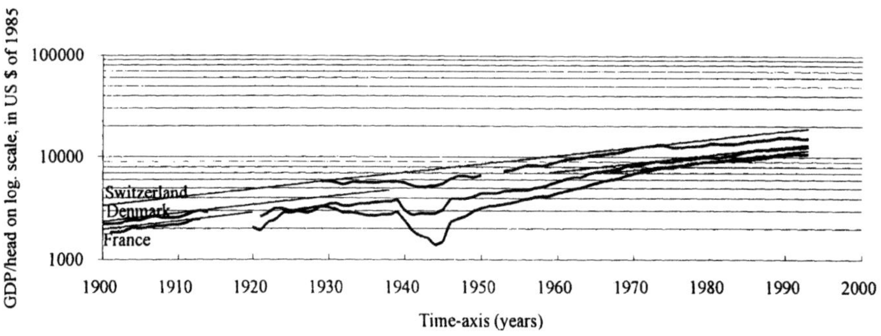

Figure 5. Development of France, Denmark and Switzerland in 20th Century

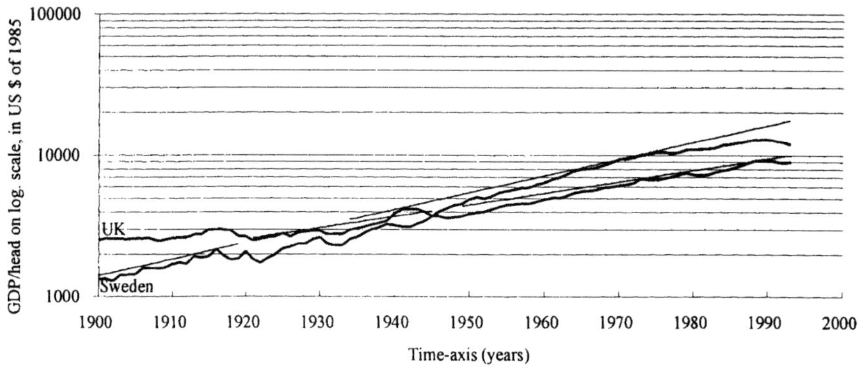

Figure 6. Development of United Kingdom and Sweden in 20th Century

of growth in the broken trendline after the 1968 break-point is 1.44 per cent. Since the difference is out of the margin of error, we can draw the conclusion that the US 20th century trendline was slightly broken by the last third of the century.

Jánossy starts his 1966 book [2] with the so-called 'Japanese economic miracle' since, until then, Japan had produced the perfectly classical traits of a reconstruction period. In his article [4], he laid more stress on the German case since, during the five years which had elapsed since 1966, forecasts for Japan had largely been refuted by the facts.

Looking at the 30 years which followed, we can safely declare that Japan is a classical example of the reconstruction period. As Figure 1 shows, the broken line with one break-point fitted to the OECD data breaks in 1970, but true recession began only later, in 1973. Jánossy's theorem is therefore (within a 3.5 per cent error margin) fulfilled.

For *Austria, Denmark and Italy* (see Figures 3 and 4) Jánossy's theorem held true with great precision. Austria and Denmark fulfilled it with a 3 per cent and Italy a 7 per

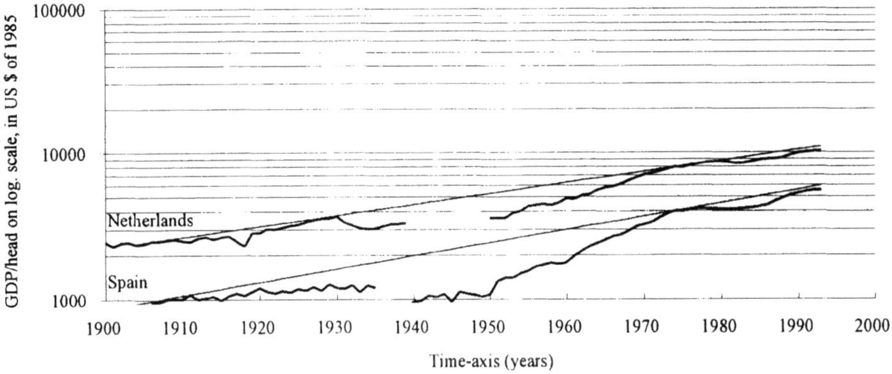

Figure 7. Development of Spain and Netherlands in 20th Century

cent error margin. In France, the GFR and the United Kingdom, Jánossy's theorem holds true with error margins of 8, 3 and 4 per cent respectively. In the remaining five countries (the Netherlands, Norway, Spain, Sweden and Switzerland) Jánossy's theorem does not hold true.

4.5 MADDISON'S DATA

In the last column of Table 2—Jánossy-trendline for 20th century Maddison's data—the rates of growth of the Jánossy trendline are calculated from Maddison's database [7]. These data give us nearly the same picture as that of Mitchell's data. The differences are slightly greater in the latter but still within the formerly applied limit of 8 per cent, except in Italy where Mitchell's data show 11 per cent as opposed to 7 per cent on Maddison's.

5. Explanatory Proof of the Law Today

Since the empirical proof of the law holds true, even when extending the series for thirty years, the 'human capital' explanation must remain true also. To put it another way, the relation—summed up in Jánossy's five points quoted here in section 3.1—between the change in 'professional structure' of the manpower of a country and the rate of growth of economic development, remains the same. On the basis of point 5 the 'new technology' can influence the economic progress of the less developed countries only through their 'professional structures'. The barriers (point 3), which limit the speed of change of the 'professional structure' also limit the long-run economic development rate. Therefore the barriers which limit the speed of change of the 'professional structure', at the same time limit the transfer of new technology.

6. Conclusions

In summary then, we can see that all a government's actions and efforts to extend capacity barriers limiting the speed ot change of the 'professional structure', extend, at the same time, the barrier to technology transfer. On the basis of Jánossy's theory, therefore, these type of investments into human capital prove likely to be the most important barriers to technology transfer.

Notes

1 This chapter relies largely on the results published in [5] and [15] which were obtained through a European Union's COST (A7) research project entitled, 'The role of some long range regulators in the integration process of the East Central European countries toward a single European market'. The project was carried out by J. Kovács, G. Molnár, T. Tarján and I. Virág at the Institute of Economics, Hungarian Academy of Sciences, from 1993 to 1995.

2 See [2] and [3]. We note here that Jánossy had predicted, at least five years earlier, the stagflation of the 1970s and in spite of the fact that his theory had a positive acceptance, nobody believed him. To make it clearer let me quote for example Samuelson's [2] textbook, page 853: 'Then the 1970s ushered in the "age of stagflation", in which rising inflation and unemployment were to appear simultaneously as the growth in living standards slowed sharply. This development was on no scholar's timetable—not seen in the crystal ball of Spengler, Toynbee, Marx, Schumpeter, or Galbraith. We live in a world no prophet ever predicted!'

3 The project was entitled 'The role of some long range regulators in the integration process of the East Central European countries toward a single European market'.

4 Since nearly all Jánossy's book [2] is devoted to this problem we must be satisfied here with summing up his multi-faceted and accurate proofs without aiming at completeness.

References

[1] Denison, E. F. (1985), *Trends in American Economic Growth, 1929–1982*, Brooking, Washington, D. C., and in U.S. Department of Labor, 'Multifactor Productivity Measures, 1986,' October, 1987.

[2] Jánossy, F. (1966a), *The Trendline of Economic Development and Reconstruction Periods* (in Hungarian), KJK, Budapest.

[3] Jánossy, F. (1966b), *Das Ende der Wirtschaftswunder: Erscheinung und Wesen der wirt-schaftlichen Entwicklung*, Neue Kritik, Frankfurt.

[4] Jánossy, F. (1971), 'The trendline revisited' (in Hungarian), *Közgazdasági Szemle,* 7–8, pp. 841–867.

[5] Kovács, J. (ed.) (1995), *Technological Lag and Intellectual Background: Problems of Transition in East Central Europe*, Dartmouth Publishing Company, Aldershot, and Brookfield USA.

[6] Liesner, T. (1985), *Economic Statistics 1900–1983*, The Economist Publications Ltd, London.

[7] Maddison, A. (1989), *The Word Economy in the 20th Century*, OECD Development Centre, Paris.

[8] Mitchell, B. R. (1975), *European Historical Statistics 1750–1970,* Macmillan, London.

[9] Molnár, G. and Tarján, T. (1995), 'Productivity lag and intellectual background', in Kovács, J. (ed.), *Technological Lag and Intellectual Background: Problems of Transition in East Central Europe,* Dartmouth Publishing Company, Aldershot, and Brookfield USA, pp. 85–107.

[10] OECD (1993), *National accounts 1960–1991*, Main aggregates,OECD Publications, Paris.

[11] Ray, G. F. (1991), *Innovation and Technology in Eastern Europe. An International Comparison*, National Institute of Economic and Social Research, Report Series, no. 2. London, pp. 123.

[12] Samuelson, P. A. and Nordhaus, W. D. (1989), *Economics* (13th edn), McGraw-Hill Book Company, New York, St. Louis.

[13] Sipos, A. (1995), 'Technology development in Hungarian agriculture: results and problems', in Kovács, J. (ed.), *Technological Lag and Intellectual Background: Problems of Transition in East Central Europe*, Dartmouth Publishing Company, Aldershot, and Brookfield, USA, pp. 31–4.

[14] Tarján, T.G. (1991), 'L'innovazione tecnologica nell'economia ungherese', Atti del Convegno su "La Nuova Ungheria e i Rapporti Internazionali", *I.S.I.G. Working Papers,* pp. 19–35.

[15] Tarján, T.G. (1995), 'Imminent OECD membership of Hungary and the revival of Jánossy's trendline theory', *Acta Oeconomica*, 47 (1–2), pp. 111–136.

TECHNOLOGY TRANSFER IN THE TRANSITION ECONOMIES: CASE STUDIES

BARRIERS TO TECHNOLOGY TRANSFER IN CENTRAL AND EASTERN EUROPE

KATALIN BALÁZS
Institute of Economics, HAS
1112 Budapest Budaörsi út 45
Hungary

1. Introduction

The following chapters highlight some of the characteristics of central and east European research and innovation systems. This region has been in transition from central planning to a market economy since 1990 and its political and economic restructuring have created uncertainty and strengthened the barriers to technology transfer. Innovation and technology transfer are not key issues in the transition period at all. Thus when we look at these particular issues in order to understand changes, we should start by understanding the background from which these countries began to move. These chapters give some insight into different countries' innovation systems and their changes during the transition period. Two of them (Egorov and Sandu) are focused on their research systems and changes in general, while the Bulgarian (Todorov) tackles the specific issue of academic entrepreneurs.

2. Research System

There is a common historical structure in central and east European research systems. The institutional complex developed in the USSR was introduced in the other countries, along with the common ideological focus on the political role of science in 'scientific socialism'. The design of the R&D system followed the general principles of central planning. It was developed using a top-down process in the spirit of 'specialisation', 'rationalisation' and 'centralisation' around distinct sectors: the academies, the universities and the branch or industrial institutes.

The National Academies for Sciences were organised to carry out basic research into the main disciplines. The assemblies of the academies had a role as a society of honoured scientists (like the Royal Society in the UK), and at the same time were responsible for science policy and coordination of basic research throughout the country. These dual roles had been overlapping and confusing and even conflicting. The assembly members were selected on the basis of honour and reputation or political connections, but not for their managerial and policymaking skills. The institution was extremely hierarchical, with power in the hands of a narrow elite and the bureaucracy. In order to become an

J. Kirkland (ed.), Barriers to International Technology Transfer, 149–155.

academician, a researcher had to pass in sequence four different research degrees. These were, the Doctor of the University, Candidature (these two degrees have since been merged to form the PhD), Doctor of Science, and Appointment to the Academy; for each degree the researcher received a supplementary salary from the central government budget. In role and social status, academicians were a kind of *nomenklatura*. In the original model, every new academician received a research institute, over which he or she had direct control (within the limits of state central funding, which was decided in response to an Academy plan). In the Soviet Union this led to a multiplying of institutes with considerable 'ballast', and a tendency to scientific conservatism. Elsewhere the model was not followed so strictly and the management role gradually moved away from academicians. Indeed, one of the most striking features of the transition was the conflict between the academicians of the General Assembly and the managers of the institutions.

Universities were devoted exclusively to education rather than research. Higher education and research were divided from each other but competed for the same budget source so that an unhealthy rivalry developed between the two systems over the years. However, in spite of poorer physical assets and funding, some research activity has developed over time in the university sector. For example, lecturers working on their research degree carried out research with their students. As a result of this activity, the level of university research at the time of transformation was by no means negligible.

Applied research and development was organised in industrial research institutes under the auspices of 'branch' ministries. (In the 1950s industrial branches each acquired their own ministries.) These were also budget running institutes working under plan for the firms of their branch. Later their role varied from country to country. In Hungary the industrial research institutes were transformed into state owned R&D enterprises. In Romania they played the key role in the R&D system in the period of extreme autarchy of the 1980s [10].

During the very strict central planning period the enterprise was seen as a simple executive body. Thus research activity was abolished and shifted either to industrial or academic research institutes. Due to the basic needs of production, particularly in branches such as pharmaceuticals, it slowly began to reappear on a small scale. For this reason in-house R&D still plays a small role in east European industry. To explain the non-innovative character of the shortage economy it is necessary to examine other factors.

There was no relevant high tech industry in the region (despite the defence sector in the Soviet Union) so the industrial partners' needs had to come from traditional sectors and matured technologies—from manufacturing units which saw no real need for R&D. However they were aware of needing help to improve both bottlenecks in manufacturing, technical breakdowns and quality problems with raw materials or components from suppliers. Local engineers were often not sufficiently up-to-date to deal with technical problems. So although the research sector was designed to supply advanced scientific results for any industrial sector, most of the firms had, in reality, entirely different needs.

In a market economy management is responsible for market strategy and incremental innovation. However, under central planning technical development had been managed through large discrete investments. Moreover, in Hungarian economic literature only investment was considered to be an economic factor, while technical development was left out and consigned to the attention of engineers and technical staff. The firm as an executive body might be better understood as a manufacturing unit without a management function. Thus between the discrete investments, the firm continued to run new technology and improved solutions to emerging technical problems, but there was no incentive and equally no capability to incrementally upgrade technology. Thus not only was the in-house R&D capacity missing, as many studies on eastern Europe have pointed out [21], but the management function as well. The impact on the relationship with the research sector was that the content of the 'research contract' could not grow out of any 'innovation strategy' but rather from everyday technical problems. This restricted 'executive' characteristic of the firm explains why it was so difficult to bring in any new technology alongside the main investments, and why the firm was such a poor learning organisation as well as teaching object.

These potential industrial partners defined the character of academy–industry relations. The relationship and its impact on the business activity of the research institutes and universities has been analysed elsewhere [1], describing also typical 'quasi-entrepreneurial' activities, like industrial contracts (under these conditions), in-house production and services. Here we will focus on the content of these activities in order to learn about capabilities accumulated through them.

There were some similarities between the central planning and the western innovation model in that early period (for example, in the 1960s). Both systems considered science to be a leading force and provided increasing public investment for it. The disseminated Soviet science system based on the linear innovation model is best understood as a linear technical development model. (In central planning the term 'innovation' excludes marketing and services as well as business interest.) The chains of the linear model were organised in separate institutions, each with a clear, simple purpose and organisation, sharply separated from each other [1].

The R&D system was fragmented. Thus a gap has developed not only between academia and industry but among all kinds of institutions; between academic research and universities, basic and applied research and so on. Funding went to institutions rather than to individuals or research groups; evaluation was weak and formal. A wide range of research fields was also typical due to the autarchic approach. This is one reason why priority setting is so difficult in the transition period.

3. Country Differences

Although, due to economic and political changes over the last two decades, the general institutional framework is rather similar across the region, as are the main structural problems, differences have developed among several countries. In Hungary the new economic mechanism and a consistent, pragmatic political approach from 1968 onwards introduced a market style of behaviour, although the basic structure remained unchanged, as did the institutional system of R&D. However, there was an impact on science policy. In Czechoslovakia in 1968 an entirely different change took place and the system again became strictly centralised and politically controlled. In Poland economic problems led to political dictatorship in the 1980s with falling living standards and public expenditure and an exodus of intellectuals. By the 1980s, when the Romanian system reached its crazy stage, with an extreme political dictator in a fully closed economy, basic research was abolished and the academic research institutes thus lost significance. However, at the same time industrial research had been strengthened in order to be able to copy and adapt technology for import substitution.

These differences had an impact on the R&D system. It was either modified (as in Romania, where it shifted towards industrial research [10]), or adjusted (as in Hungary) to contract research and develop business interests [1]. In addition there was growing pressure on the R&D institutions caused by a lack of funds. To combat this, research groups in many countries developed income-making strategies. Signs of production and contract research could also be discovered in Czechoslovakia, Poland and Germany.

Looking at the level of funding over time and the role of qualitative evaluation gives us more information on the present circumstances of domestic R&D. Analysis of the impact of funding and evaluation is not in any case easy. Data collection is not equivalent to OECD standards and thus represents only tendencies. The funding level was rather high; in the 1960s and 1970s generally it was close to 3 per cent of GDP. The timing of dramatic drops in funding varied from country to country; for example in Czechoslovakia it remained high until 1989 (over 3 per cent of GDP) and has been falling rapidly ever since. In Hungary the R&D system in the 1980s was already suffering from decline. Cuts in expenditure have been forcing adjustment for more than a decade and the decrease in the last three years has been even sharper [2] [5].

4. The Transition Period

The economic transition has been accompanied by deep economic crisis. Political changes have broken formal economic links and the east European market has collapsed. The crisis has spread from country to country and from branch to branch. The firms most affected were those specialising in CMEA exports. Industrial production fell dramatically in line with the fall in GDP, the most critical period being between 1989 and 1993. In the worst years the fall in industrial production reached a critical –25 to –30 per cent and GDP has fallen by an average 20 per cent over three years. In some countries there were some signs of recovery by 1993 (Poland, Hungary), but in the countries of the former Soviet Union the economic crisis has been longer and deeper, observers reporting that the Russian economic situation is chaotic and lacking perspective.

The transition itself has had side effects since new legislation set new conditions and while generating restructuring it also gave rise to uncertainty and distorted former production and market links [31]. Firms were governed only by short-term interests and most rapidly therefore cut out any R&D expenditure. The formerly high level of R&D expenditure thus fell below 1 per cent in two or three years. Although the effect was dramatic, the impact had different effects on different organisations. Typically, industrial R&D suffered most, due to shrinking markets and disappearing demand.

Transition policy in most countries has followed the fashion of focusing on monetary and fiscal policy, but it has not been complemented by economic foresight and long-term strategy. Separate policies have pursued particular objectives. The mission of privatisation was to increase the private sector, while that of tight monetary policy was to control inflation and balance the budget. As a result of high-interest loans, investment became very expensive. Firms found themselves in difficulties, and many faced bankruptcy due to a lack of working capital. Capital shortage and expensive loans, together with concentration on short-term interests, are the reasons behind the falling investment rate. Economic policies continued to limit domestic markets and import liberalisation opened up the market to foreign competitors. The almost total elimination of price subsidies has made exporting more difficult and has increased competition.

As a result of declining expenditure and a changing economic environment, the R&D sector has experienced a passive structural change [3]. A structural shift has occurred in the R&D sector because industrial expenditure fell more than the the state budget. Industrial research suffered most, while academic research has turned away from technology application. The R&D sector has not been in the forefront of economic policy and so has not figured as even part of short-term interests. Competitiveness as a policy target has not yet emerged anywhere. The R&D sector has been treated as any other inherited problem, with its centre as part of the 'old' bureaucracy. Thus science and technology policymaking has been lumped together with other policies and other government interests. It has all been a rather slow and conflicting process. Moreover, there is a long history of disagreement among the organisations involved. The R&D sector and its bureaucratic centres have suffered from institutionally built-in tension. Traditionally the Academy confronted the universities, and the ministries each other. Thus their position in lobbying for a coordinated government strategy was rather weak.

In these years, therefore, R&D systems have suffered from general, though uneven, losses in terms of funding and numbers of researchers, so the key issue in the transition lay with priority setting and evaluation as a main instrument for restructuring. Qualitative selection has been made neither by economic needs and market forces nor by competition for grants. Funding of institutes was still typical and cutbacks affected the average financial level. A National Research Fund at the project level was first created in Hungary in 1984 as an attempt to give a greater role to evaluation and scientific competition and other countries set up their research funds in the early 1990s.

In the policy vacuum individuals and research organisations rather than policymakers have developed active adjustment strategies. On the one hand, research managers and new technology-oriented entrepreneurs have formulated alliances, associations and societies as pressure groups bridging the gaps between each other and between academy and industry. These associations have tried to substitute science and technology policy and stimulate policymakers to act. On the other hand, research institutes, along with individuals, have carried out local organisational reforms and developed new practices to gain a better living. These attempts led to the formulation of science parks, technology centres and spin-off companies, all based on growing entrepreneurship [3].

References

[1] Balázs, K. (1993), 'Lessons from an economy with limited market functions : R&D in Hungary in the 1980s', *Research Policy*, 22, pp. 537–552.

[2] Balázs, K. (1994), 'Transition crisis in Hungary's R&D sector', *Economic Systems*, 18, 3, September, pp. 281–306.

[3] Balázs, K. (1995), 'Innovation potential embodied in research organisations in Central and Eastern Europe', *Social Studies of Science*, 25, 4, November, pp. 655–685.

[4] Balázs, K., Falkner, W. and Schimank, U. (1995), 'Transformation of the research systems of post-communist Central and Eastern Europe: an introduction', *Social Studies of Science*, 25, 4, special issue, November, pp. 613–633.

[5] Balázs, K., Hare, P. and Oakey, R. (1990), 'The management of R&D in Hungary at the end of the 1980s', *Soviet Studies*, 42, 4, pp. 723–741.

[6] Cave, J. (1991), 'Political reform and scientific freedom under Gorbachev', *Technology in Society,* 13, 1/2.

[7] Darvas, Gy. (1988), *Science and Technology in Eastern Europe*, Longmans.

[8] Fortescue, S. (1990), *Science Policy in the Soviet Union*, New York, Routledge.

[9] Ionescu-Sisesti, I. (1992), 'Transition and conversion toward innovation in Romania: Factors and actions', Paper prepared for the IACHEI Annual Conference, September 22–25.

[10] Ionescu-Sisesti, I. and Sandu, S. (1993), 'The innovation paradigm —utility and possibility for Romania, *Costinesti*, 4–5 June.

[11] Jasinski, A. (1991), 'Recent changes in the Polish R&D system', *Science and Public Policy,* 18, 4, August, pp. 245-251.

[12] Jasinski, A. (1994), 'R&D and innovation in Poland in the transition period', *Economic Systems,* 18, 2, pp. 117–141.

[13] Josephson, P. R. (1994), 'Russian scientific institutions: internationalisation, democracy and dispersion', *Minerva*, 32, 1.1,Spring.

[14] Kornai, J. (1980), *Economics of Shortage,* Amsterdam, North Holland.

[15] Lenardic, M. (1992), 'Science and technology policies and economies in transition: a case study of Croatia, mimeo.

[16] Loudin, J. (1992), *Problems of Science and Technology Policy in Society in Transition*, Centre for STS Studies, Prague.

[17] Lubrano, L. (1991), 'New initiatives and old bureaucrats: the future challenges the past in academic science', *Technology in Society*, 13, 1-2, pp. 91–109.

[18] Lubrano, L L. and Solomon, S. G. (eds) (1980), *The Social Context of Soviet Science*, Boulder, CO: Westminster Press; Folkstone, Kent: Dawson.

[19] Meske, W. (1993), 'The restructuring of the East German research system —a provisional appraisal, *Science and Public Policy*, 20, 5.

[20] Müller, M. (1992), 'Academy–industry relations in CSFR: an attempt to assess a historically founded perspective, mimeo.

[21] Pavitt, K. and Hanson, (1987), 'The comparative economics of research development and innovation in East and West: a survey' in Montiars, J.M. and Kornai, J. (eds) (1987), Harwood Academic Publishers.

[22] Piskunov, D. and Saltikov, B, (1992), 'Transforming the basic structures and operating mechanisms of Soviet Science', *Science and Public Policy*, 19, 2, April, pp. 111–119.

[23] Radosevic, S. (1994), 'Strategic technology policy for Eastern Europe', *Economic Systems*, 18, 2, pp. 87–117.

[24] Radosevic, S. (1995), 'The Eastern European latecomer firm and technology transfer: from "muddling through" to "catching up"', in Alferov, Z., Bugliarello, G. and Pak, V. (eds), *Technology Transfer: Personnel and Leadership Issues*, Kluwer Publishers.

[25] Ray, G. (1991), 'Innovation and productivity in Eastern Europe: an international comparison, *National Institute Economic Review*, 183, pp.75–84.

[26] *Review of National Science and Technology Policy: Czech and Slovak Federal Republic* (1992), Paris, OECD.

[27] *Review of National Science and Technology Policy: Hungary* (1992), Paris, OECD.

[28] Schimank, U. (1995), 'Transformation of the research systems in Central and Eastern Europe: a coincidence of opportunities and trouble', *Social Studies of Science*, special issue, November.

[29] Simeonova, K. (1993), 'Conflicts in Bulgarian scientific system in the transition to market economy', *Science Studies*, 6, 1, pp.196–208.

[30] Simeonova, K. (1995), 'Radical and defensive strategies in the democratisation of the Bulgarian Academy of Sciences', *Social Studies of Science*, special issue, November.

[31] Swaan, W. (1994), 'Behavioural constraints and the creation of markets in post-socialist countries', paper given at conference in Trento, Italy, March.

[32] Török, A. (1991), *Market Orientation of Hungarian Enterprises -EC Integration*, Research Institute of Industrial Economics, Budapest, October.

[33] Vlaicu, C. (1993), *Academic–Industry Relations. A Romanian Case—A Soiological Point of View*, Science Technology Center, Bucharest University.

[34] Weiss, C. Jr. (1993), 'The re-emergence of Eastern European science and technology', *Technology in Society*, 15, pp. 3–23.

OBSTACLES FACED BY SMALL FIRMS IN THE TECHNOLOGY TRANSFER PROCESS IN ROMANIA[1]

STELIANA SANDU
Institute of National Economy
Bucharest

1. Introduction

Before 1989 any concern over technology transfer in Romania was limited to the diffusion and generalisation of new or improved technological procedures and processes obtained within departmental research institutes. These Institutes were set up between 1970 and 1989 to carry out research and design for thousands of enterprises in 'multilateral developed' industry. Far too many large institutes were set up at this time and, by the end of 1989, there were 343 R&D units, most of them industrial research institutes.

Imports of technology and equipment were stopped after 1982. There was talk in Romania of 'the extension and generalisation of the technical progress', and about 'world-level and over world level products', but these were only slogans [8]. While the large industrial, agricultural and construction enterprises predominated, productivity in the SME sector was almost non-existent. Now, however, there are concerns at both a theoretical and strategic decision-making level regarding the development of SMEs, in particular of those which are innovatory and absorb new technologies. Statistical data for 1994 show 67 per cent of SMEs in services (mainly commercial), 25 per cent in industry, 6 per cent in agriculture and 2 per cent in construction.

2. The Economic Background to Technology Transfer in Romania

The progress of economic reform in Romania has been influenced both by historical factors and by factors specific to the transition period [12]. The main features of the economy before 1990 were as follows.

- A national production system with numerous oversized productive segments and low technological and economic efficiency, compared with the available material resources.
- The capability of these oversized segments to adjust to the dynamics of the real restructuring processes and to the new requirements of the market proved to be extremely low since they represented the most rigid components of natural production.
- Low labour efficiency in most economic sectors as a direct effect of oversized employment, the low use of production capacity, slack labour discipline and the lack of labour incentives.
- A distorted system of maintaining unprofitable enterprises, through redistribution of financial resources provided by profitable units at both national and sectoral levels,

J. Kirkland (ed.), Barriers to International Technology Transfer, 157–170.

has had negative consequences on the supply side [14].

- The extremely harmful effects, at a structural level, of the foreign debt advance pay-off policy. This had the effect of forcing economic activity while resource allocation was increasingly uneconomic. Excessive exports triggered a drastic decrease of domestic final consumption while the severe cutback in imports strongly diminished competition in technology—the most productive and innovative characteristic of the economy.

Under these circumstances, evolution of the national economy against a background of deepening structural imbalances generated substantial growth in the money supply while the country was increasingly incapable of backing up its national currency with goods and services [3]. Consequently, even stronger signs of repressed inflation growth were seen, reflected, particularly after 1985, in the severe shortage of consumer goods and in the drastic cutbacks in investment in production. These imbalances have put pressure on the functional structures of the Romanian economy since 1989, particularly in view of growing inflationary tendencies which the liberalisation processes of economic reform have triggered.

It has become clear that the hasty dismissal of the mechanisms of a centralised economy has failed to lead, as anticipated, to the establishment of a market economy. The mentality, behaviour and industrial structure specific to the former system persisted up to 1994, with inadequate management of assets, poor capitalisation of resources, consumption growth of raw materials and low labour motivation. The situation has worsened because some enterprise managers have raised prices after giving in to higher wage claims, although production has decreased.

The process of economic reform was begun in 1989. A first phase saw dynamic legislative and institutional change together with small steps towards economic stabilisation. In the second phase efforts were directed towards ensuring and consolidating macroeconomic stabilisation, while institutional change continued at a slower pace. The aim was to ensure a stable economy by removing major imbalances (high inflation, a growing currency deficit) through the use of tight money, tax, credit and income policies. The economic situation in 1993 (the end of the second phase of reform) is given in table 1, using the main macroeconomic indices.

The reform process proceeded against a background in which economic mechanisms were neither of a socialist economy nor of a market economy. Economic policy was at first based on the idea of classical liberalism but eventually proved to be incompatible with the realities of transition. The process of reform and the pressure of specific problems forced the government to assume responsibility by resorting to economic intervention.

Problems of high inflation and foreign shortages were fuelled by the gap between consumption and available resources resulting from the previous policy of import substitution together with forced and accelerated industrialisation. This led to an excessive diversification of the branch structure and to a lack of international specialisation which limited the possibilities for export.

There has been only a partial and temporary adjustment to consumption through control of aggregate demand. The slowdown of excessive diversification in the branch structure of industry and success in overcoming the lack of international specialisati

TABLE 1. 1985–1993 Synthetic indicators (1980=100, per cent)

Indicators	1985	1986	1987	1988	1989	1990	1991	1992	1993
GDP	116.8	119.6	120.6	120.0	113.0	106.7	92.0	83.6	84.7
GDP by economic sector, of which:									
Industry(a)	82.2	86.8	92.2	93.2	85.4	71.1	61.9	55.7	58.0
Construction	117.0	121.5	122.9	119.7	95.9	96.9	78.1	72.4	85.0
Agriculture & forestry	115.7	110.0	105.0	110.0	103.3	144.0	126.9	106.4	119.2
Transport	109.4	115.5	120.3	119.8	116.1	90.8	75.5	70.6	57.0
Trade(b)	102.3	100.1	100.8	108.4	113.0	119.9	93.9	77.6	78.3
Post & communications	154.6	126.3	130.0	110.0	98.1	79.7	110.2	115.6	121.1
Finance, banking & insurance	140.7	138.7	240.4	213.4	104.9	123.1	122.7	167.0	171.6
Immovable dealings & other services	112.9	117.5	120.8	124.5	114.4	118.9	116.5	135.9	118.5
Public administration & defence, mandatory social assistance	171.1	106.6	111.6	104.5	113.6	129.4	119.1	119.7	125.9
Education	107.6	103.6	100.2	99.6	103.1	108.7	126.6	126.2	137.6
Health & social assistance	118.7	117.7	117.7	116.9	128.9	160.3	157.1	145.3	139.4
GDP per capita	114.1	116.3	116.7	115.5	108.4	102.1	89.0	81.4	82.6

Source: *Romanian Statistical Yearbook*, 1994, National Commission for Statistics, p.361.
(a) Including electric and thermal energy, gas and water. (b) Including hotels and restaurants.

goals which could not be attained through macroeconomic policy alone. They imply effective restructuring in the real economy, especially in industry.

Development of the private sector has taken place according to the Land Law and the Law of Privatisation. About 80 per cent of total arable land has been returned to its original landowners but law enforcement is a complex process, given the necessity of accommodating subsequent regulations, especially those concerning railway traffic [15].

The law on companies' privatisation regulates the transfer of state property to the private sector, represented either by natural or legal persons, Romanian or foreign citizens. On the one hand, the law stipulates distribution of free vouchers to Romanian nationals, when procedures are applied to commercial companies, and when there is participation of natural or legal persons, Romanian or foreign citizens, to the sale of shares or assets pertaining to commercial companies. On the other hand, the law stipulates the setting up and the structure of certain institutes directly involved in the restructuring process: these are the State Ownership Fund, five private ownership funds, and the National Agency for Privatisation. By 1994, only 8 per cent of commercial companies had been privatised. Their capital share represented only 3 per cent of the total of those commercial companies in line for privatisation.

TABLE 2. The main indicators of aggregate supply in 1994, per cent

	1994/1993
Gross domestic product (GDP)	103.5
Industrial production	103.3
Construction	118.3
Agriculture	100.2
Transport	96.4

Source: National Commission for Statistics

TABLE 3. The main indicators of aggregate demand in 1994, per cent

	1994/1993
Final consumption:	104.7
- household	105.3
- public administration	101.7
- private administration	100.0
Gross fixed capital formation	115.9
Exports of goods and services	125.8
Imports of goods and services	110.8

Source: National Commission for Statistics

2.1 THE THIRD STAGE OF THE REFORM PROCESS

The 1993 trend towards revival of production was consolidated in 1994. As may be seen from table 2, growth occurred in the main aggregate demand indices, that is GDP, industrial, agricultural and construction production.

The statistical data show a significant contribution from industry and construction as opposed to the negative contribution of services. As regards aggregate demand in 1994 compared with 1993, there was significant growth in fixed-asset formation (115.9 per cent), as well as in exports of goods and services (125.8 per cent), compared with final consumption (104.7 per cent). The inflation rate has slowed down considerably, from 12.1 per cent in 1993 to 4.1 per cent in 1994, and the average depreciation rate of the exchange rate has fallen from 5.8 per cent in 1993 to 2.8 per cent in 1994.

The annual average growth rate of net nominal wage income decreased from 9.9 per cent in 1993 to 5.8 per cent in 1994, highlighting a fall in living standards. By the end of 1994, the unemployment rate was 10.9 per cent.

Property structure has changed owing to the acceleration of the privatisation process of state owned companies and also as a consequence of the development of small and medium-sized enterprises. Loyal competition has been promoted and protected. Autonomous state-owned enterprises are being restructured, and the process of restructuring and readjustment of industrial production has been intensified. Consequently, unprofitable

TABLE 4. The share of the private sector, per cent

	1993	1994
Gross domestic product	32.0	35.0
Industrial value added	14.4	15.0
Value of construction work	35.1	57.9
Cereals production	79.9	82.5
Retail sales	55.1	68.8
Fixed investment	26.0	38.4
Value added in trade, hotels, restaurants	56.8	69.8

Source: C.N.S.

industrial production has decreased. Economic agents are being stimulated with a view to increasing export oriented production. Management quality has improved and so has financial discipline. The position of the private sector within the economy grew stronger in 1994 compared with 1993, as shown in table 4.

The obvious discrepancy between the rise in the number of private-capital companies and the growth in the share of the private sector in GDP indicators points to the existence of significant difficulties in the development of private firms. The significant increase of the private sector share in retail sales suggests that its expansion has been accelerated more as regards GDP utilisation than GDP creation and the great difficulties which have been faced in production activities.

Since 1995 the mass privatisation process has been initiated by changing social capital into shares. Thus 3,000 enterprises will be privatised and 60 per cent of their social capital will be converted into shares. The number of profit-oriented private R&D institutes has also increased; about twenty R&D institutes, mainly involved in design, were privatised through the MEBO (management employees buy-out) method and the R&D unit privatisation process is to continue until the end of the year. Up to fifty R&D units are currently operating as commercial companies.

Privatisation of R&D institutes has raised debate and controversy amongst those involved in S&T policy in Romania (such as the Ministry of Research and Technology, the National Union of the R&D) and design institutes, as well as within each institute, amongst different interest groups. The differences in opinion arose mostly with respect to the assessment of the intellectual assets pertaining to these institutes, when they were left out of consideration in the general legislation, in terms of intellectual and industrial property rights [11]. In accordance with some legal provision these institutes have been given real estate, second-hand equipment and buildings for their own use.

Strategy was lacking in the selection of the R&D institutes to be privatised. The government was indiscriminate, voluntarily turning half the technological research institutes (about 100) into commercial companies, basically profit-oriented. Thus, from 1991, many research institutes became commercial companies, carrying out profit-making activity, as well as research. A major way of securing income was in some cases by renting the space leased to small or large companies (tens of thousands of which were set up after 1990), since there was a tremendous demand for real estate. The 'lease' payment would be masked by research contracts concluded between the institutes and their lessees, which, given their specific type of activity, needed no research output, except for the leased premises [7]. Under these circumstances, some of the highly skilled researchers would desert the institute, either to set up their own private research and design companies or to emigrate (15 per cent of the total number of researchers in 1992), or to join other fields of activity, funded from the state budget or by private capital.

3. The Creation and Development of SMEs in Romania

3.1 GOVERNMENT STRATEGY

Since 1993, the government has paid special attention to the development of SMEs, setting out a special policy with the following basic objectives:

- free or low-cost information, assistance and consulting services for SMEs
- subsidies for the set-up of a national network of information, assistance and consulting centres for SMEs
- direct financial support for the development of production activities, construction, services and tourism, and agriculture
- integral or partial financing of some professional training programmes
- state allocation for the purchase of production space, tools and equipment and technology
- application of an efficient purchasing system, with the use of assets becoming available following the restructuring of state-owned enterprises
- priorities to credit lines extended by international banks and financial institutions and to government credit lines
- credit guarantees
- investment allocations
- financing of some technological innovation and research activities (including technology transfer from research to SMEs)

Some of these strategic targets have been reached, some not. Overall, however, the development process of some innovative SMEs has proved to be difficult and slow.

3.2 DIFFICULTIES ARISING FROM ECONOMIC POLICIES FOR INNOVATIVE SME DEVELOPMENT

From October 1990 until 1993 Romania experienced high and volatile rates of inflation. In 1994 there was a cut in the rate but it has continued to be relatively high for the requirements of a business-friendly macroeconomic environment, including an innovative SME sector. There was therefore a high degree of risk and uncertainty and short-term planning in trade and services and in economic agreements, and an increase in production costs which was unfavourable for productive SMEs [2].

The government solution of granting low interest credit to all SMEs, irrespective of field of activity, ignores the fact that they are affecting the stability of the financial system and generating losses in the economy as a whole. A long-term solution might be to cut inflation and lower interest rates with the aim of encouraging SMEs to invest in innovative activities. At the same time the government should promote an R&D incentive tax-credit and a wide range of linked measures to encourage the dissemination of R&D results towards SMEs.

There are many difficulties concerning the process of market creation and consolidation, such as delays in the adoption of the necessary regulations and non-correlation of measures aiming to get markets operational. All are barriers facing the development of innovative SMEs.

On the monetary side, banks are hesitating to meet the requirements of SMEs. Various problems exist—the low qualifications of staff, an overestimated level of guarantees and reserve towards granting long-term credit. Many components of the capital market are poorly developed in Romania (for example, long-term financing institutions, a secondary market of bank guarantees, the risk-capital market, leasing, factoring). This has made it difficult for the SMEs to find long-term financial sources.

The under-capitalisation of banks was a major factor leading to a shortage of SMEs' long-term financing during 1990–4. Technology transfer of course needs strategic vision and long-term investment. Difficulties generated by the slow pace of restructuring and privatisation, like the decrease in importance of asset sales within the strategy of privatisation which began in 1993, were important impediments to the development of the private sector. Government Decree no. 500/1994 on the revaluation of state enterprises' fixed assets has artificially increased their price (by between two and ten times), and has discouraged SMEs from buying shares. At the macro and regional level also, the programme for industrial restructuring has not been correlated with those intended to promote SMEs and this has lessened the SMEs' chances of developing profitable industrial activities.

Many problems have arisen in relation to the process of redefining the role of the governmental authority in the economy. Hesitation in this respect has determined the flimsiness of the measures to support SMEs and therefore hindered their expansion. Difficulties are apparent also in obtaining space for development, together with a lack of management consultancy and training. The current education system pays little attention to skills essential for the build-up of enterprise. An information system with reference to research findings and product-quality standards is also absent. The government therefore must develop programmes focussing on particulars in order to meet the specific perceived requirements of SMEs.

3.3 LEGISLATIVE AND PROCEDURAL DIFFICULTIES

The legislative system is not yet homogeneous. Many laws and regulations issued under previous regimes now coexist but often contradict each other. The adoption of a fundamentally new legislative system is a long-term process. Parliament has to adopt hundreds of laws, all of which require debate, so that some projects have to wait years before being put into practice (the Research Law has waited four years). Other contradictions arise between the stipulations of acts and the instructions for their enforcement, because of the lack of experience in the legislative framework of a market economy. Since this framework is unstable, about 38 new acts regulating the activity of SMEs were issued every month in1994, as well as the existing 112 acts from 1991–2. Many other problems slowed down the set-up and development of SMEs in fields favourable for the implementation of new technologies. Consequently there was a rapid increase of SMEs in services, especially in commerce and other profit-making services.

Some of the legislative difficulties with the setting up of SMEs are as follows:

- problems with the logistics of law enforcement
- the establishment of property rights (especially over production sites)
- the process of establishment or termination of SMEs
- taxation regulations and the relationship with the taxation authorities
- accounting regulations
- procedural difficulties in relations with the banking system and intermediary financial institutions
- labour terms (ending and breaking of work contracts, labour and social protection)
- economic agreements and contracts (very old rules from 1960—law of leasing agreements unsatisfactory.

All the above mislead and frighten small-scale businessmen and discourage them from starting up in business.

3.4 INSTITUTIONAL OBSTACLES

There is a wide diversity of institutions competing for representation. At the heart of the dispute lies the handling of funds from national and international bodies, which are earmarked for the stimulation of SMEs. For example, the CRIMM Foundation benefited from PHARE funds for establishing business incubators in different counties of Romania. On the other hand, the Ministry of Research and Technology benefited from various government funds with a view to setting up business and innovation centres and centres for invention implementation. This second type of centre provides facilities and consultancy services from professors and researchers, as well as access to research labs or to the available spaces existing within the universities or the research institutes. At present, many institutions are involved in the development of SMEs. They have parallel, uncoordinated activities, which lead to inefficient fund utilisation which confuses the small entrepreneurs. There are three governmental bodies, the Romanian Development Agency (RDA), the Council for Coordination of Reform, and the Ministry of Research and Technology (MRT). There is also CRIMM, a non-governmental body, the local public-administration bodies—the Chambers for Trade and Industry, the SMEs development centres established with government support (37 in April 1995), the business incubators, the centres for innovation and business, as well as the pilot incubators, the consulting companies, the financial institutions, the international community and the business environment representative bodies.

3.5 OBSTACLES ARISING FROM THE OFFER OF TECHNOLOGY

Between 1991 and 1995 the results of industrial research have been too insignificant to arouse the interests of small firms. Industrial research institutes were financed mainly from a special R&D fund, which drew on the 1 per cent tax on the turnover of both state-owned and private economic agents. This fund has been supervised by the MRT, who allocated funds for research not according to the needs of industrial restructuring and modernisation, but according to tradition and the institutes' individual capabilities. In these circumstances, many institutions have inserted traditional or old topics into their research programmes, which are uninteresting from the point of view of industry.

In the opinion of several MRT experts, the Special Fund for R&D has sometimes worked in the same way as the Fund for New Technique, used prior to 1989 by the centrally-planned economy, and only 16 per cent of all the themes completed by the end of 1993 were effectively transferred to industry [5].

The research financing system in Romania, reshaped in the rush to 'save' technological research, has ignored several essential principles which had provided the basis for the reform of the funding system. These principles are as follows:

- fund allocation according to the criteria of scientific value and practical use of the projects
- free access of all researchers to public funds (the Academy researchers were denied access to the 1 per cent Special Fund)
- transparency in managing public funds and strict control of the way they are spent
- the use of certain specific 'schemes' for financing meant to stimulate joint research and technology transfer, which would lead to the increase of the private sector's contribution to research financing
- promotion of various income-generating activities (in addition to the public fund), in order not to harm the quality of the research activity but to ensure a more efficient capitalisation of financing at institutional and regional level.

For Romania the most important issue is to recover the formal and informal relations between research institutes and industrial companies. Since these were suddenly terminated in 1990 this process will take time and their evolution depends on how a real competitive market will provide incentives for technology demand. The Romanian market economy is unresponsive to technological change and thus cannot stimulate innovation. 'The offer of technology' must overcome the present inertia, reconsider old research topics and supply some new solutions for industry, which is now itself subject to restructuring. It also involves government's determination to balance the research share in the public purse and to identify the best allocation of mechanisms and criteria.

The budget allocation for the national research programmes focused on the traditional branches of the processing industry (50 per cent), thus affecting certain fields which are a priority for science. For instance, biotechnology, a basic field of research and technology, has suffered a fall in its budget from 2.9 per cent in 1991 to 1.1 per cent in 1993, and this also occurred in information technology, with a drop from 4 per cent to 0.5 per cent.

The quality of the human input involved in Romanian R&D activity is another restrictive factor since it appears that young graduates are not attracted to such activity. According to the most recent statistical data, the number of employees in the field aged under 30 with university backgrounds dropped from 4,913 in 1991 to 4,406 in 1993. Similarly, the numbers of employees aged between 30 and 39 were reduced from 11,372 to 10,983 (*Romanian Statistical Yearbook*, 1994, p.314). There was also a reduction in the number of researchers (19,900 to 16,188, 1991–3) and an increase in technicians (13,826 to 16,221).

Before1995 there were several dysfunctions and gaps in Romania between different types of research institutes (technological research, the Romanian Academy's institutes, universities, industry). Unlike the situation in the developed countries, there was therefore little experience of collaborative research, which is a main prerequisite for technology transfer [4]. Although some direct contacts are made between economic agents and the research institutes, they are usually formed on the basis of traditional, often personal, links, and do not address essential aspects of research and technology transfer, but offer consulting for privatisation.

4. Small Innovative Firms in the Academic Environment[2]

According to Romania's 1994 *Statistical Yearbook*, there were 146 technological R&D institutes in 1989, 47 of which were of design for investment projects and 93 for research institutes for agricultural production. Many of the technological research institutes were equipped as workshops or microproduction plants for the production of prototypes. They also produced goods for sale to supplement their insufficient funds.

Although the research institutes served the enterprises, the contracts were drawn up with the National Council for Science and Technology, a bureaucratic party organisation. These contracts had to prove formally the economic efficiency of implementing research results in terms of production and profit growth, cost reduction and so on. In such circumstances enterprises and research institutes resorted to exaggerated and improbable reports of technology transfer. The bureaucratic activity involved in this has transformed the syntagma at the root of the theory and practice of technology transfer in the west—the education/research/production relationship—into a simple slogan [6]. Such organisational defects have reinforced habits and practices unchanged since 1989, while academic research was carried out within a framework of 'compulsory norms' in existing research institutes or enterprises and the Romanian Academy's network of research institutes was broken up. (Payment of one third of a professor's salary was conditional on obtaining a report from the beneficiary of his research specifying the fulfilment of such 'norms' [9]).

For Romania then, as for other eastern European countries, the adaptation of the R&D system to market forces took place in a tense, often confrontational, domestic and international situation. In particular, difficulties were incurred with the discrepancy between economic and social reform and the reform of the R&D system; non-existent research priorities with limited funding; the payment of compulsory1 per cent contributions by economic operators to the Special Fund for R&D Financing without any benefit from the research results; and a lack of cooperation between the government bodies set up to manage the 'national innovation system'. In contrast with other countries though, there are weak, informal relations between R&D institutes, the universities and the Romanian Academy [1].

There is a lack of public finance to ensure stability in the development of innovation and a lack also of institutional relationships to link the components of the innovation system [10]. Government in Romania has not yet perceived the necessity for change from a linear to a fundamentally different interactive model of innovation. Decision-makers have no clear perception of the advantages of feedback throughout the innovative process and of the interactions between science, technology and innovation processes. In the world at large, however, in the face of growing international competition, there is cooperation between industries in research and technology improvement through technology exchange, mixed R&D, direct investment in R&D and so on.

In 1979, within-firm research was almost destroyed in Romania with the transfer of research teams from enterprises to big departmental research institutes. Although steps were taken towards redressing this problem, a drastic capital shortage in most of the industrial enterprises had an important effect on Romanian technological development.

In brief, the present national innovation system is much as it was before 1989, linear and inefficient and stimulating neither to the demand nor the supply of technology. Relations between the parts involved in the process of technology transfer are fragmented and their reconstruction will be slow even without a national strategy backed by sufficient funding. (In addition the home market provides no incentive to innovation since consumers, having very low incomes, accept outdated low-quality products.

A good indicator of the innovative propensity of a country is provided by statistics on invention licensing. The value of such indicators in Romania over the last five years has steadily fallen, despite the fact that several prizes have been won at international invention fairs and there is now also an entire network of centres for invention implementation. (For example, in 1990, the total number of licence requests was 3,081, falling to 1,826 in 1993, of which only 1,414 came from nationals.)

Researchers studying the relations between academics and industry have stressed the importance of government backing for such activity through the Ministry for Research and Technology (MRT). In 1992 a representative of the MRT had the idea of setting up two institutional networks for facilitating technology transfer—the innovation and business centres' network and the invention implementation centres' network. The rapid creation of these two national technology transfer networks within a year is surprising at a time when research activity was fumbling towards a market economy [1].

Since 1990, when several academics became involved in entrepreneurial activity, Romania has learned from the experience of other countries, thus stimulating and facilitating the establishment of small innovative companies, usually of the 'spin-off' type. The main types of spin-off in this environment are:

- SMEs set up by professors or researchers wishing to trade their own research results
- SMEs set up by graduates wishing to exploit the results of research in which they had been involved
- SMEs set up by outsiders to the academic environment intending to use the results of research carried out in universities or research institutes.

The establishment of these new types of innovative SMEs was a combination of bottom-up and top-down policies. First the National Agency for Privatisation and then the MRT gave financial and logistic support for up to three years but the results subsequently obtained often failed to live up to initial hopes and enthusiasm. Researchers, stimulated by the funds allocated by the MRT for incubation and business centres, and by the availability of premises and equipment within the institutes, have continued their initial researches (where these had commercial possibilities) in their spare time. Their aim was not to set up new SMEs, but to benefit from the facilities offered by the Institute in order to undertake research with an assured market (for example, ZECASIN S.A., Bucuresti; The Centre for Invention Implementation, Craiova; The Centre for Technology, Invention and Business, Brasov). Thus government funding was the main force behind the proliferation of innovation and business centres in the three years to 1995 when, according to MRT data, there were thirty such centres.

Although most of these centres have also undertaken research in areas related to those for which they have received public funding, the continuation of their support has been justified by the fresh approach they have brought to the academic and research environment and also by some good results, particularly from the Craiova and Brasov centres. Regretfully, public funding of these centres has now been stopped, owing to financial irregularities discovered by the authorities between the working status of the centres and present legislation.

5. Concluding Remarks

Technology transfer was only slowly introduced in Romania between 1990 and 1994 because of difficulties inherent in the transition to a market economy and a lack of any coherent government policy. The SME sector was less involved in the process since there was a lack of economic incentive and of an institutional and legal framework available to sustain the requirements of SMEs for new technologies. Lack of government incentives in the form of regulations has provided an unsuitable environment for innovation and there is consequently a low rate of domestic and foreign technology dissemination. Research collaboration between enterprises is almost non-existent, though the Academy's institutes are at least engaged in relationships with enterprises or other research institutes. Funds for research are not allocated on a priority basis as this exists neither at R&D nor at industry level.

Governmental or non-governmental bodies might assist SMEs to develop strategies for technology transfer, linking them effectively into their overall business plans. Government can play an important role in stimulating a collaborative approach to R&D either directly through sponsorship of support schemes or by other actions. It could develop policies to increase the efficiency of the existing research system, including the development of guidelines for the planning and management of research programmes, their assessment and evaluation, and the development of guidance sources for SMEs on how best to utilise R&D as part of a wider business strategy.

From 1995, some measures have been taken to solve some of the problems mentioned above. In May 1995 a government resolution (H.G.294) was made regarding the stimulation of economic agents in the process of technological transfer of the results of publicly-financed R&D. The White Paper on The Romanian Academy Research Strategy also came out, with four key objectives including competitive financing and integration of Academy research with University research, and a forthcoming abrogation of some legislation which had discouraged Academy institutes from having research contracts with economic agents. Technological research is to be oriented towards the needs of end-users, changing the financing system of big public research institutes so that they will be financed through grants based on programmes initiated and approved by government.

Notes

1 The research project underlying this chapter is part of an ACE-PHARE Program 1994 funded fellowship project (94-0446-F): Technology Transfer from Research to Industry.

2 The Institute of National Economy has obtained information about the experience in this field of other countries through their collaboration in the international project: Innovation potential embodied in the changing Academy–industry relations in east Europe. We would like to express our gratitude to the coordinators of this project, Katalin Balázs, Andrew Webster, Slavo Rodosevici and Peter Healy.

References

[1] Balázs, K. and Plonski, G.A. (1994), 'Academic–industry relations in middle countries: East Europe and Ibero-America', *Science and Public Policy*, April.

[2] CRIMM (1995), *White Paper of the SMEs*, The Foundation 'Romanian Center for Small and Medium-sized Enterprises'.

[3] Croitoru, L. (1993), *Macrostabilization and Transition*, Bucharest, Expert Publishing House.

[4] Faulkner, W. (1992), 'Understanding industry–academic research linkages: towards an appropriate conceptualisation and methodology', Working paper no. 35.

[5] Georgescu, A. and Bastiurea, Gh. (1994), 'Strategia MCTG in domeniul valorificarii rezultatelor activitatii de C-D', paper presented at a seminar on 'Technology Transfer, Incubators and Innovation Centers', Bucharest, December.

[6] Gibbons, M. *et al.* (1995), *The New Production of Knowledge*, London, Sage Publications, pp. 46–57.

[7] Ionescu-Sisesti, I. (1992), *Transition and Conversion towards Innovation in Romania, Factors and Actions,* Proceedings of the IACHEI Annual Conference.

[8] Ionescu-Sisesti, I. (1994), *A Case Study of Reform of the National Research*, Washington DC, George Washington University.

[9] Ionescu-Sisesti, I. and Sandu, S. (1993), *Science and Technology Policy in Romania during the Transition Period to Market Economy*, Bucharest, p. 27.

[10] Niosi, J. and Bellon, B. (1994), 'The global interdependence of national innovation systems, evidence, limits and implication', *Technology in Society*, 16, 2, pp. 173–97.

[11] Sandu, S. (1992), *Study on Technological Research Institutes' Privatization*, Information Bulletin 4, Bucharest, Ministry of Education and Science.

[12] Sandu, S. (1995), 'Innovation potential embodied in the changing research–industry relation in Romania', *Romanian Economic Research Observer*, 2, 5.

[13] Stanciulescu, I. and Prodan, Gh. (1994), 'Internationalizarea tehnologiilor, transferul acestora si implicatiile asupra evolutiei societatii romanesti', Bucharest, Economic Advisory Group.

[14] Zaman, Gh. (1994), 'Restructuring state enterprises in Romania', *Romanian Economic Observer*, 2, p 45.

[15] Zaman, Gh. (1995), 'Economic dimension of sustainable agriculture', *Romanian*

Economic Review, 1–2, pp. 9–32.
[16] Zaman, Gh. and Sandu, S. (1994), 'Considérations préliminaires concernant le fonctionnement de l'économie nationale dans les conditions du socialisme monopoliste d'Etat', in *L'économie de la Roumanie – le XXe siècle*, Bucharest, Les Editions Expert, pp. 407–83.

TECHNOLOGY TRANSFER IN UKRAINE:
Slow Changes on a Background of Fast Economic Decline[1]

IGOR EGOROV
Centre for R&D Potential Studies,
National Academy of Sciences of Ukraine
60 Shevchenko boul., 252032, Kiev, Ukraine

1. Ukrainian R&D and Innovation Systems in the late 1980s and early 1990s

The Ukrainian economy and its systems of innovation were formed as integral parts of the Soviet Union, with a higher level of integration into the overall structure than that of eastern Europe. Present Ukrainian economic and innovation systems are therefore inherited from that period.

Three main features should be mentioned with regard to the R&D system. First, technological development has grown up largely in isolation from the rest of the world. This has influenced the way in which R&D specialists solve problems—often without reference to economic reasoning—particularly in the short term. At the same time, as a result of this isolation from the scientific and technological community, some original and effective solutions have been found.[2] Second, science has habitually neglected market influences. Science and technology appear to have been more imposed on society in the Soviet-type economies than in the west. Third, a serious problem arose because of the branch structure of monopoly, preventing technological transfer between industries and enterprises.

The Soviet economy was characterised by profound structural and technological imbalances. As Schneider [20] noted, 'both the substantial discrepancy between administrative management methods and the innovation processes, and the economic, ideological, and socio-cultural peculiarities have resulted in an economic and political indifference to innovation and technological change.'

The use of ideological regulators of economic development and the existing systems of planning and evaluation of scientific and technological activity resulted in a unique exaggeration of the quality and quantity of results, and a trend toward spectacular projects with sensational results.

As Varshavsky [23] showed, existing practices of innovation evaluation and implementation of arbitrary coefficients in standard methods of efficiency calculations have led to huge overestimation of the real results of innovation process for the R&D institutes and, at the same time, to their underestimation for industrial enterprises. It seems that present and future costs of innovating would be far higher than potential returns and an enterprise would be better off adhering to traditional technology.

J. Kirkland (ed.), Barriers to International Technology Transfer, 171–184.

The closed nature of the centrally planned system and its commitment to secrecy imposed additional extensive limitations on the mobility of skilled personnel and the transfer of information and technology. At the same time, there was pressure to follow western countries in civilian production partly to meet the growing consumer demands of the population. In conditions of self-reliance, 're-invention' and copying of western-made products became widespread in eastern-bloc countries [6] [11]. The practice of centralised purchasing of imported equipment, with its chronic delays in supplies of spare parts and tendency to save currency on service operations, also contributed to the importance of the economic function of domestic S&T.

International R&D cooperation is vital for both domestic technological progress and competition, which reflect the evolution of demand in modern society. Reverse engineering was one of the few links of the Soviet specialists to the international S&T community. This peculiar research method was particularly popular in the motor vehicle, aircraft, chemical, and microelectronics industries. It was engaged in simply to organise production and not for long-term improvement or development technology. Although there was some trade in licences, a high proportion of Soviet technology imports were merely of the turnkey style that limited further domestic development and expansion. The direct import of knowledge and exchange between established or aspiring scientists was minimal.

Glaziev and Schnider [8] noted that branch or industrial research institutions were simply extensions of the ministry bureaucracy. Each ministry or branch had its own R&D establishment; each branch, and indeed each institute or *nauchno -proyzvodstvennoe ob'yedinenie* (scientific and industrial complex), tried to be self-sufficient. R&D institutions responded to pressure from the hierarchy (upstream), and were relatively cut off from the users of their products (downstream).

Attention to the downstream phase of innovation activity was rarely important in meeting efficiency requirements. In conditions of artificial pricing for many products in the Soviet economy, it was possible to establish extremely low costs for inflow products such as colour metals and special alloys. That is why assessment of the effectiveness of new technologies in the Soviet Union revealed contradictory results with respect to contemporary global trends.

Concepts of organisation, planning, and management were borrowed mainly from material production. Technological progress in industry was defined as an increase in unit capacities of existing equipment types, often beyond economically (and ecologically) justifiable limits [19].

Ukrainian R&D potential held a special place in the former Soviet Union. The Ukrainian science establishment was huge. The Academy of Sciences comprised about 180 institutes; 160 universities with research departments were situated in Ukraine; more than a thousand institutions in branch industries concentrated their efforts in specific sectors. Under the Soviet Union, Ukraine held 23 per cent of all invention certificates.

Ukrainian R&D became famous above all for its achievements in the area of technology. Even the Academy of Sciences had a 'technological' orientation. But at the same time Ukrainian specialists did not take an active part in international cooperation, and this sometimes led to a failure to understand trends in world scientific and technological development and a tendency to overestimate their own results in R&D. In the review of priorities of Ukrainian science conducted in 1989–90, the absolute majority of scientists declared that their R&D work met or surpassed world standards (!)[16]. But the facts do not support this assertion.

The majority of R&D institutions, and indeed of industries as a whole, were administered by ministries located in Moscow and usually had very few contacts with local enterprises. Thus, in Ukraine in the late 1980s, more than two thirds of all orders for R&D came from other parts of the Soviet Union [21].

Exports from Ukraine were to and through the USSR. Most contacts with the outside world were through Moscow. There was little need for Ukrainian officials and specialists to become familiar with such issues as marketing, promotion, intellectual property protection, R&D strategy, logistics and negotiations. These were handled by the central government in Moscow and its specialised organizations.

The political changes provoked by *perestroyka* have created a new economic environment. Parallel with the move towards democratisation, enterprises received some freedom in their economic activity. Most of them, being monopolist in particular areas, did not behave in accordance with the hopes of Gorbachev's advisers. Instead of allocating resources to new technologies and equipment, they started to increase the level of salaries of their employees, without real growth of productivity. The creation of new forms of enterprises, such as Young Communist League Centres for Promotion of S&T Activity and cooperatives, made the situation even worse, because they opened up new channels for turning money from the accounts of state companies into cash [12].[3] These things, along with the partial legalisation of 'gray' and 'black' economies, were the initial cause of serious imbalance in the financial system, and later led to the collapse of the Soviet economic and political system.

Given the highly monopolised internal market and weakening control on the part of the Ministries, enterprises stopped making orders for R&D, and stopped payments to centralised funds for S&T development. Even if there was a need for some R&D work, enterprises preferred to make deals with cooperatives or small groups of scientists. In this way they could obtain the same product without having to pay the overheads of big institutes, because the legal regime in the former Soviet Union with respect to intellectual property rights was underdeveloped, and specialists from the R&D sector could usually use the equipment and results of their institutes without proper control. These developments were accompanied by the depletion of centralised funds for S&T activity, and a decrease in the salaries of many specialists in their primary place of work.

2. After Gaining Independence

Until 1991 the Ukrainian economy was part of the Soviet economic system. But in 1989–91, it was doing better than the Russian economy for a number of reasons [14].[4] The trends in the R&D sector were also similar, but the destructive processes in 1991–2 were slower than in Russia [13].

The proclamation of independence encouraged a hope that the Ukrainian R&D sector would break its dependence on Moscow and start to forge its own development path [15] [18]. In reality, the opposite happened. The breaking up of the Soviet Union left many institutes without most of their customers, without marketing or negotiation skills, and without adequate intellectual property protection, capital and other tools needed to survive over a long period.

This can be explained partly by the fact that the majority of the users of Ukrainian S&T output remained in the other republics of the former USSR. Permanent disputes with Russia and economic crisis throughout the former USSR led to a decline in the level of cooperation. In 1992 only 11.7 per cent[5] of all financing for R&D projects in the Ukraine was obtained from abroad, mainly from the other CIS countries. According to Professor Lev Gassanov,[6] who was the head of the military–industrial conglomerate 'Saturn', the Russian government established strict terms for the reorientation of their enterprises towards domestic suppliers of R&D output. Such measures cut off the Ukrainian R&D sector from its most important markets.

Ukrainian leaders had also counted on strong domestic demand for R&D output, but the economic situation made this unrealistic. Since gaining independence in 1991, the Ukrainian economy has performed worse than any other in eastern Europe except Serbia, which was under an international blockade.[7]

Two main 'objective' features of R&D itself prevent it from playing a more active role in the process of economic transformation.

1) There is the uncertainty of final R&D results. Few are prepared to allocate money to R&D in an unstable economic environment. The production sector has anyway enough financial difficulties. Only trading operations remain profitable.[8]

2) The R&D cycle is comparatively long. Even where results are predictable, net profit could be negative, because of high inflation and other forms of instability in the economic 'environment'. At the state level, R&D has no positive impact on short-term stabilisation, which interests government and parliament much more than the strategic issues of economic development.

Three-quarters of all enterprises have already stopped any inventive activity and activity connected with rationalisation. Only 5 per cent of all R&D results obtained in 1991–3 met world standards and only 0.5 per cent of enterprise production was competitive in the international markets in 1992–3. Less than 12 per cent of all engineering production was exported in 1993, but only 2 per cent of it for hard currency.

The 'jumping' character of inflation in the Ukraine makes it impossible to include accurate information on the R&D sector. In 1992 alone, prices changed rapidly twice, in 1993 another four times, plus 'ordinary', 'smooth', inflation; and in 1994 twice. At the same time adjustments to the funding of R&D institutions were substantially delayed. A special study would be required to assess absolute financial trends with accuracy. Preliminary analysis shows that even the nominal volume of funds for R&D work rose at only half the rate of inflation in 1992–3.

A further decline in innovative activity was observed in 1994, when the values of all indicators connected with innovative activity and technology transfer were negative. So, industry-oriented R&D fell by more than double during 1991–4 and in some key sectors of the economy the decline was even more acute: in the construction industry it fell to a fifth of the previous level; in machine-building to a quarter; and in colour metallurgy to a third.

The decline in R&D in most progressive energy and material saving technologies was even greater, particularly in metallurgy, and the wood and timber industry [22]. These sectors can now support only small incremental innovations.

Due to the rapid decline in innovation and investment flows, the rate of renewal of equipment in industry became lower than 1 per cent a year (in respect to its total value). The contribution of innovative activity into total enterprise income has fallen by a third. Approximately the same picture may be observed in productivity growth dynamics: the rate of productivity growth, connected with the implementation of new technologies, declined from 2.2–2.3 per cent in the 1980s to 0.7 per cent in 1994. Some hopes arose with the commercialisation of the R&D sector and creation of private innovative companies and cooperatives, which in 1989–92 represented 12–18 per cent of all the enterprises of this type.[9] But these enterprises were usually relatively weak; the number of employees was several times less than in enterprises in other sectors, and the share of part-time employees exceeded 90 per cent [4]. In 1993–4 the share of private R&D companies and cooperatives declined significantly from 2.4 per cent to nearly 0 per cent, whilst the so-called 'mixed' (semi-state) companies experienced significant growth—from 2.8 per cent to 4.3 per cent.

For some firms 'science' has become a shield, which helps them to avoid high taxes and, of course, the R&D sector takes an active part in 'wild' privatisation, when equipment and buildings of state-owned institutes and design bureaux are used to obtain private profit without any payments to the state budget. 'Mixed' companies are vivid examples of this tendency. Such activity helps people survive but, unfortunately, it has nothing to do with R&D. Usually these new firms are more involved in trade and other types of commercial activity.

Bearing in mind the misuse of existing S&T potential, it seems that the plans of the Ukrainian government to make its country self-sufficient in R&D will not be realised. The main actors in internal (Ukrainian) and CIS innovative processes remain the same: state R&D institutes and state enterprises. In conditions of severe economic crisis, it would not be possible to stimulate innovative activity without effective state support.

Hopes that Ukraine could be included as a more active participant in international technological cooperation, through multinational operations, joint ventures or licensing agreements, have also been unfulfilled [3].

Few of the joint ventures which have been developed have involved R&D. As Goldman [9] stressed, the more immediate demand for venture partners has been for the production of, for example, expensive cars, rather than for long-term research and development. This situation persists, despite statements from managers and directors of joint ventures, to promote promising innovation.[10] Unfortunately, the main purpose behind the creation of joint ventures by western partners is often just to penetrate the new market and obtain short-term benefits. FDI could play an important role in technology transfer, but its influence in the Ukrainian economy to date is negligible.

Precise data is lacking, but it is clear that few large and well-known corporations have joint ventures with partners from the former Soviet Union. With a few notable exceptions—such as Johnson and Johnson—joint venture partners have concentrated on the sale of manufactured goods produced elsewhere, rather than developing new production lines.

3. Institutional Changes and Market-Oriented Innovative Capabilities

Institutional problems have been mentioned above. The process of innovation suffers from continuing regulation by the former Soviet laws. In recent years, however, the process of change has begun.

Legislation concerning science and technology policy passed by the Ukrainian parliament in 1991 laid down basic principles for the functioning of an independent R&D sector. It also opened the way for later legislation to clarify different aspects of R&D activity. 1993 saw new laws concerning the 'protection of rights on inventions and useful models', the 'protection of rights on prototypes' and on 'protection of brand names and trademarks and copyright protection'. These steps slightly improved the situation with regard to the regulation of property rights. In 1994, the Ukrainian budget received more than $2 million from intellectual property rights protection by foreign inventors. At the time of writing, further legislation on science and technology expertise was under preparation.

The situation concerning patenting rights remains complex. Following the collapse of the former Soviet Union, few Ukrainian scientific institutes and design bureaux were granted patent rights, especially in the west. The majority of existing patents were inherited by Russian bodies, as the leading or main institutes in particular sectors. The lack of foreign patents owned by Ukrainian entities is a tremendous liability in negotiations with foreign companies, and the marketing of high-technology products abroad.

The Ukrainian Patent Office provides information on how to file internationally and selection of patent attorneys. This has limited effect since inventors lack the funds to implement their advice. This includes not only finance to file patents, but to pay maintenance fees. There is, however, significant interest in domestic patenting. Between 1991 and 1994 over 50,000 applications were filed for Ukrainian patents, more than 90 per cent of them from domestic inventors. Unfortunately, the state patent office does not have the resources to keep pace with this demand. By mid-1995, 2,500 patents had been granted and about 3,000 trademarks registered.

This situation places Ukrainian institutes and specialists at a disadvantage in their negotiations with western partners. By not adequately protecting Ukrainian interests, it ensures that technology and know-how is sold to western interests on highly unfavourable terms [2].

One example of this was the sale, by a cybernetics institute, of patent rights to new information storage methods for CD Roms. The initial payment of $25,000, for exclusive marketing rights, did not even repay the institute's investment in patenting fees. Further royalties on sales were not forthcoming when the purchaser, recognising the technology as a potential rival, declined to develop it.

The state patent office seriously lacks infrastructure. It has no database of inventions and applications, and no computer network to handle its functions. There is little information on past patents, making the search process difficult. Thus the loss of intellectual property rights is common. One example of this is the case mentioned above, where some scientists leave their institutes, taking potentially valuable research results with them, and fail to compensate the laboratory concerned. Thus the institutes receive no royalties for some of the research which they have funded.

Where royalties are being earned, a high proportion is often paid to intermediaries. Some small US brokers have received 50 per cent of royalty earnings, chiefly for filing patents and finding partners. A more normal fee in such cases would be 10–30 per cent, depending on the difficulty of the task and the value of the product.

In summary, Ukraine suffers from the lack of a general system for intellectual property protection. The country has joined the patent cooperation treaty, the Paris convention on industrial property and the Madrid convention on trademarks. But domestic legislation for regulating R&D is not complete, and Ukrainian institutes have neither the funds nor the skill to defend against violation of their intellectual property rights abroad through securing rights, monitoring use, or hiring appropriate legal representation.

To illustrate this situation in practice, we now turn to three typical cases of innovation and technology transfer.

4. The Case of the Aviation Industry[11]

The Ukrainian aviation industry comprises over 50 enterprises. In 1989, it employed over 150,000 workers, a figure which had fallen by 30 per cent by mid-1995. Three enterprises may be considered as critical to the industry; the Antonov aviation complex, the helicopter factory and the Motor-Sich engine factory. The industry was initially part of the aviation capacity of the Soviet Union. Thus, the level of cooperation in technology and production was exceptionally high. Historically, 85–90 per cent of the industry has been deployed in Russia, and other republics had to rely heavily on supplies from that source.[12] There was little capacity for distinct R&D activity in the Ukraine.

Ukraine occupied a special place in the Soviet aviation industry, largely due to engine production (in Zaporozhie) for several types of Soviet planes, and to the Antonov aviation complex, which was noted for its design of cargo planes. The Antonov complex was especially important, since it provided R&D for a family of AN planes in Russian aviation factories. Following the collapse of the Soviet Union, cooperation between Ukrainian and Russian enterprises became complex. Ukrainian enterprises lost their former leadership in composite materials and other fields. Leading producers, such as the Illyshin and Typolev complexes, preferred to equip their planes with British or American engines to meet western standards for noise and environmental protection. Russian enterprises have no special leaning towards Ukrainian partners, but some factories in Russia still needed technological supervision from the Antonov bureau, and preferred to use engines from 'Motor Sich'.

One of the more successful examples of technological cooperation between the two countries concerned the creation of the AN–70 multi-purpose cargo plane. Work on this plane began in 1987, with a view to delivering a new product, capable of replacing the relatively obsolete AN–74 and entering the world market by 1991. The design stage incorporated several new innovations; the plan could be used either as a cargo or passenger plant, with capacity for 52 persons; it was capable of being converted between use within hours and aimed for a speed of 700 kilometres per hour and flight distance of 4,500 kilometres.

The project was a genuine joint venture in its finance and production. 8 per cent of funding came from the Russian defence ministry, and the balance from the Ukrainian ministry for machine building and conversion. 75 per cent of all components were received from Russia but the engines were made in the Motor Sich plant.[13] Good personal relationships between officials in the Russian and Ukrainian aviation industries played a key role in the project's development. The first plane was sold to a Siberian gold-mining company prior even to the first flight test programme.

The project continues to demonstrate great potential, with the prospect of orders for at least 500 planes and employment for 27,000 highly-qualified staff in Ukraine, and some 50,000 in Russia. It is even possible that the finished product will compete with the new European cargo plane FLA, which is currently at the design stage. However political changes and economic decline have led to numerous delays, and the plane will need 250 checking flights and approximately two years of other tests prior to commercialisation. The desire to commence flights as soon as possible led to a catastrophic trial for the first-assembled plane in Spring 1995. At the time of writing, it is not clear whether good personal relations and enthusiasm will be sufficient to take the project through to fruition.

Another project to generate optimism was the production of a new AN–180 plane, designed at the Antonov bureau. The specification was for 150–180 passengers, 3000 kilometre flight distance and efficient, low-noise engines from Motor-Sich which would meet western standards.

Design work on the project began in 1993. According to the head of the Antonov complex, Peter Balabyev, the main problems have been unsuitable financial regulations and a lack of state orders.[14] Although the Ministry for Machine Building and Conversion has provided 20–30 per cent of the required budget, the remainder has to be attracted from commerce. The designers have some expectation of Russian demand, although this may be difficult to fulfill.

Finally, some interest has been expressed in the use of Ukrainian planes outside the CIS. The US air force has investigated the possibility of combining heavy Ruslan planes with Pratt and Whitney engines and electronics from General Electric, while Federal Express intended to test the AN–38. In each case, some characteristics of the planes need to be significantly improved, requiring the Ukrainian aviation industry to transfer know-how from outside the country. For the reasons expressed above, there are few signs of such transfer taking place. Most of the results now being utilised are based on R&D work completed in the 1980s; aviation enterprises have preserved only part of their potential and have been further hampered by the 'brain drain' of staff, accounting for over 40 per cent of key personnel between 1989 and 1995.

In these conditions four main scenarios are possible for the future development of the industry. First, it could seek to maintain its independence. Such an approach is unlikely to be successful, however, because of the economic difficulties and technological gaps identified above, which prevent an autonomous Ukrainian industry competing in the world market. A second scenario involves a transformation of the Ukrainian aviation industry into small companies, focusing on cargo flights and technical assistance to countries where links already exist, particularly the former Soviet Union, India and Poland. This might provide some future for the industry, whilst not addressing many fundamental problems.

A third scenario involves reconstruction of relations amongst CIS producers. Some special forms of cooperation have already been proposed for this purpose including the creation of so-called international industrial–finance groups. The first group of this kind was created in the aviation industry, but met with some resistance in the Ukrainian parliament in early 1995. Nonetheless, measures of this nature could help in the reconstruction of in-house R&D, the importance of which to growth in high-tech industries has been described above.[15] Such development could also help avoid excessive taxes and customs.

The fourth scenario involves extended cooperation with western partners. Thus far, the industry has not received any significant investment from western companies, many of whom would not wish to encourage the emergence of a new competitor. But some proposals for cooperation have been mooted, and Fokker expressed an interest in joint production.There is also potential for Ukrainian companies to participate internationally as parts suppliers [24]. Such projects would help enable Ukrainian enterprsies to raise their technological and marketing capabilities, but psychological and institutional barriers thus far have prevented their development.

The ideal solution will probably lie in a combination of the last three scenarios, with particular emphasis on the third and fourth. The first alternative would lead to further decline, since without intensive technological exchange it will prove impossible to compete even in domestic markets.

5. The Case of the Defence Industry

Many of the above characteristics can also be seen in the defence industry. In the 1980s, over 75 per cent of all R&D was devoted to military purposes; now the bulk of military orders have disappeared. For example, the Arsenal optical enterprise, once leading its field in the former USSR, has seen the proportion of its production for military purposes decline from 86 per cent to 2 per cent.

Again, the nature of this decline can largely be explained by trends in Ukrainian–Russian relations. In 1989, 79 per cent of total Soviet final arms production was produced in Russia [5]. Ukraine suffers not only from a dependence on Russia, but from the absence of a full technological cycle on its territory. Thus, for example, although Ukraine contributed work for the former Soviet navy, major science and technology centres for the purpose were based in Russia. The Ukrainian shipbuilding industry therefore lacks its own R&D base.

Having lost their traditional partners in the Baltic region and Ukraine, Russian military enterprises are now building their own full production cycles, including research and development. These developments are inevitably at the expense of Ukrainian enterprises and science and technology institutions. The history of the defence sector also inhibits present-day prospects for diffusion of military technologies. For many years, innovation in the field was characterised by dramatic technological heterogeneity. This was reinforced by the emphasis on secrecy in Soviety society as a whole. Thus, as some western specialists have already observed, the specificity of military technologies available limits the prospects for significant transfer to civil use [10].

These factors influence not only the degree of flexibility in products, but the entire value system of the workforce. Often, the combination of military and civil production in the same enterprise leads to a mixture in the form of organisation required, or the transfer of defence industry production methods to civilian products. This mixture could lead to rising costs for civilian-based production.

Some enterprises have tried to tailor their conversion to an import substitution policy. Thus, the largest space and military missiles factory in Ukraine has begun to produce trams and trolley buses for an internal market, replacing imports from Czechoslovakia. This production is still state subsidised, however, and the chances of its becoming more efficient are limited due to the high cost base of such large multi-profile enterprises [1].

Attempts to access the world market for arms export also suffer through a lack of consumers for military production and difficulties in entering a highly competitive and divided market. Two recently created companies for the marketing of Ukrainian weapons —Ukrinmash and Progress—are, however, making desperate efforts to promote new products to a world audience.[16] Ukrainian weapons attracted considerable attention at the IDEX–95 military exhibition in Spring 1995, when 158 companies expressed an interest in various products, and 56 signed pre-contract documents. Interest focused on tanks, military trucks, anti-aircraft complexes, optical equipment and the repair services of Ukrainian specialists.

The promotion represented a significant demonstration of Ukrainian potential, involving the efforts of over 30,000 scientists from 379 design bureaux and 1,700 enterprises.[17] But the participation in international exhibitions alone, and reliance solely on domestic potential, will not in itself resolve the problems of conversion to new economic conditions for the Ukrainian military–industrial complex.

6. The Case of Brody Technopark

In seeking to create an environment conducive to technology transfer and innovation, the government has sought to implement best practice from the west and Japan in developing technoparks. The first such park was created by special decree of the Council of Ministers in June 1994, based on a former Soviet missile base. Fourteen companies participated, with state support and substantial assistance from Germany and other European countries. Special conditions to attract overseas investment included tax reductions, liberal customs regulations, profit repatriation advantages and accelerated writing-off of capital assets.

Results in the first year of operation were relatively poor, especially in relation to technological development. Only three of the first fourteen participating companies could be considered technologically oriented. Limited success has been achieved only in the field of biotechnology, through a joint venture with German partners, which became the main shareholder in a plant which began to produce components for Germany and German-owned enterprises in eastern Europe. There were also some developments in the field of agricultural technology.

Although overall exports from the park exceeded $2.4 million, the bulk of these were connected with the garment and food industries. This could not, therefore, be regarded as assisting technological development. Levels of investment and of participation of foreign companies were disappointing. There have also been criticisms that foreign help was largely spent on high salaries for technopark staff, overseas visits, cars and office equipment, and that the private sector land developers have exploited the tax advantages of the site by using it for non-technological purposes such as warehousing.

Two general conclusions can be drawn from the Brody technopark experience. First, it highlights the danger of genuine attempts to promote the technology base being exploited by exclusively commercial operations, with no high-tech involvement, which are able to use regulations to their advantage. Second, the example shows that the creation of a technopark is a more organic process than was expected by the developers. It is not possible to establish the necessary conditions by decree, or with an immediate financial inflow. Other conditions, such as demand–push from industry, are essential. At the same time, lack of the necessary critical mass of qualified personnel is not compensated for by the energy of local entrepreneurs, particularly if these have no connection with the innovation process.

7. Conclusion

The specific problems of innovation systems in Ukraine are often submerged in the issues of economic stabilisation. At the same time, capacity to develop technologically advanced sectors are import preconditions for recovery and stable economic growth.

The basic problem for Ukrainian economic and technological systems is to provide sufficient demand for R&D products during the transition. Privatisation policy and selected state support could be key instruments in this process, but the monopolistic tendencies in the Ukrainian economy mean that privatisation itself will not guarantee sufficient competition to stimulate innovation. Thus, the process by which new technology-oriented companies, which invest in R&D, must be accelerated. Ultimately, it does not matter whether industrial enterprises absorb R&D laboratories, or research institutions buy factories; the important point is that they must be integrated. In-house industrial research, closely linked to the production process, is the cornerstone of R&D in a market economy.

There is need for effective economic and legal mechanisms for the acquisition and adoption of new technology, and for specific measures to support those wishing to split away from large institutions and establish small R&D companies. The state could also provide the mechanism by which some research institutes could convert into universities. Other key problems concern the cultivation, improvement and utilisation of technological capabilities within industrial companies and institutes [7].

As the process of economic transition proceeds, state subsidies should comprise a smaller share of total R&D expenditure. It is critical that these are effectively prioritised. The state committee on science and technology has already developed a list of technologies considered important for Ukraine to develop. These include health care, environmental protection, agricultural production, treatment and storage, information technology and new materials. The list identifies in detail specific areas of research related to contemporary Ukrainian problems.

Unfortunately, the process of establishing R&D policies remains heavily influenced by the political situation, and existing distribution of power, rather than a rational approach. Ukrainian authorities still rely heavily on domestic R&D and existing modes of technology transfer. Early statements from the new Ukrainian minister on science and technology problems hardly addressed the question of technology transfer at an international level.[18]

An international element is vital to the discussion. It should be remembered that technologies for domestic development and products ready for competition in the world market are different things. Ideally, a product capable of international sales should also solve a domestic problem. In this way, government support can address both needs. Officials might consider this criterion when recommending technologies for support.

At present, the import of new technologies is more important than export. Some experts argue that the more technologies which can be imported, the faster they will grow. Whilst the technological balance of payments would be negative, the trade balance could run a large surplus (as in Germany, Japan, South Korea, Taiwan and other states). Other problems, such as the need for hard currency, will be addressed in the long term, but some strategic vision from officials on R&D imports would be helpful immediately, particularly where such imports are inhibited by the financial limitations on enterprises. Finally, international advice could be useful in the technology assessment exercise itself. A first step in this exercise could involve systematic evaluation by foreign experts of research institutions and production facilities, giving independent opinion on the quality of their research and products. Such experts would be able to better assess these by reference to the quality and standards of competitive institutions, products and services elsewhere. Following the technical assesment, the next stage would be independent market assessment of particular technologies. A relatively objective procedure of this nature would provide a welcome shift from reliance on existing practice and sentiments about the 'good old days' which, along with reliance on traditional state support, may make the process of decline irreversible.

Notes

1 I would like to express my gratitude to Dr. Vladimir Karpov and Dr. Vladimir Mikhailov from the Ukrainian Institute of Statistics for allowing the use of statistical information obtained from their Institute.

2 For example, the creation of sub-gradient methods of optimisation problem solving as a result of restriction of the export of modern computers to the Soviet Union in the 1970s and 1980s and development of precise mechanics for the aviation industry as a response to a similar ban on exports of sophisticated electronics from western countries.

3 In the former Soviet Union circulation of cash and non-cash assets used in exchange between enterprises were kept separate. The volume of cash flows was subject to special control from the state financial institutions. (See, for instance, Khanin G., 1991, 'Problemy Izmerenya Natsionalnogo Dokhoda SSSR', Moscow, *Ekonomika*.)

4 See, for instance,the analysis of the dynamics of key economic indicators of the CIS

countries in *Kommersant* N11, March 29, 1994, pp. 52–3. The Soviet financial system made it possible for the Ukrainian government to export inflation to Russia, thanks to uncontrolled rouble subsidies to the 'black holes ' of the republic's economy.

5 According to various estimates (see Skhema, 1989), 50–70 per cent of all the orders for R&D from the Ukrainian R&D sector in the 1980s were from the other republics of the Soviet Union.

6 Unpublished report of seminar on 'Systems Analysis of the Transformation Process in Ukraine', March, 1993.

7 *Rossia*, March 23–29 ,1994

8 *Finansovy Kiev*, N7, 1993.

9 Ukrainian statistical publications contain a special section devoted to the performance of non-state organisations like cooperatives and small enterprises.

10 Interviews with Mr Yuri Dukhota, head of Bigarso Ukrainian–German–Swiss joint venture (biotechnology) and Dr Sergey Mischenko, head of Ilta Ukrainian–French joint venture (cars, distribution and services), 1993–4.

11 This part is based on the author's semi-structured interviews with leading specialists of the Ukrainian and Russian aviation industries, made during November, 1994–January, 1995.

12 Not only Ukraine, but Georgia and Uzbekistan had relatively big aviation enterprises.

13 *Novosty*, 22 December, 1994.

14 *Zerkalo Nedely*, 28 August, 1995.

15 *Ekonomika y Zhizn*, May 1993, N.18.

16 *Zerkalo Nedely*, 18 February, 1995.

17 *Novosty*, 13 March, 1995.

18 *Zerkalo Nedely*, 9 September, 1995.

References

[1] Bhaduri, A. and Laski, K. (1994), 'Three aspects of transformation from the command to the market economy', Paper presented to the conference on 'Alternative Ways of Economic Reform in Ukraine', Kiev, December 1–4.

[2] Brandwayn, S. and Diamond, S. (1994), 'Turning science into money in former Soviet Republics', *Law in Transition*, Winter/Spring, pp.24–27.

[3] Cusumano, M. and Elenkov, D. (1994), 'Linking international technology transfer with strategy and management: a literature commentary', *Research Policy*, 23, pp. 195–215.

[4] Egorov, I. and Solominskaya, E. (1994), 'Process Stanovlenia Novuh Organizatsionnyh Form Nauchno-Tehnicheskoy i Innovatsionnoy Deyatelnosty v Ukraine', *Reports to the XIV Kiev International Symposium on Science of Science and S&T Forecasting*, Kiev, Naukova Dumka, pp. 54–59.

[5] Falzman, V. K. (1990), 'Strukturnye preobrazovanya, NTP y ekonomicheskaya reforma', *Ekonomika y Matematycheskye Metody*, N1, pp. 67–73.

[6] Fortescue, S. (1990), *Science Policy in the Soviet Union*, London.

[7] Freeman, C. (1982), *The Economics of Industrial Innovation*, Pinter, London.
[8] Glaziev, S. and Schnider, A. (1993), 'Restructuring Russian R&D', *Options*, IIASA, Summer, pp. 12–13.
[9] Goldman, M. (1991), 'Diffusion of development: the Soviet Union', *American Economic Review*, 2, pp. 276–281.
[10] Gummet, P. (1994), 'Main trends in conversion in R&D', Report to the NATO Workshop, Budapest, August 28–31.
[11] Hanson, P. (1981),*Trade and Technology in Soviet–Western Relations*, London.
[12] Izotenko, N. (1991), 'Effektivnost Raboty Tsentrov NTTM', *Ekonomika Sovetskoy Ukrayny*, 11, p. 93.
[13] Kitova, G. A. (1994), 'The scientific potential of the former Soviet Republics: development under new conditions', *Studies on Russian Economic Development*, 5, 2, pp. 154–162.
[14] Kryzhanovsky, B. (1993), 'Yak zhivesh, Ukrayino?', *Visnyk Academyi Nauk Ukrayny*, 11, pp.13–19.
[15] Pavlovsky, M. (1992), 'Pro economichnu polityku Ukraynskoi derzhavy', *Rozbudova Derzhavy*, 4, pp. 10–13.
[16] *Prognoz Osnovnyh Napravleniy Razvityay Nauky*, (1991), 1–13, Kiev, Naukova Dumka.
[17] Radosevich, S. (1993), 'Strategic technology policy for eastern Europe', Mimeo, SPRU, University of Sussex.
[18] Romaniv, O. M. (1991), 'Nayka y problemy Ukraynskoy derzhavnosty', *Visnyk Akademyi Nauk Ukrayny*, 10, pp. 10–16.
[19] Saltykov, B. G. (1990), 'Naychny potencial SSSR: perestroyka struktury', *Economika y Matematycheskie Metody*, 1, pp. 122–134.
[20] Schneider, C. (1991), 'On research and development management in the transition to a market economy', IIASA Working Paper, WP–91–44, November.
[21] *Skhema Razvytya y Razmeschenya Otrasly Narodnogo Khozyastva SSSR Nauka y Nauchnoe Obsluzhyvanye* (1990), Moscow, GKNT.
[22] *Stan Nauki V Ukrayni u 1994* (1995), Report to the Council of Ministries of Ukraine, April.
[23] Varshavsky, A. E. (1985), 'Prognoznye modeli dlia issledovania vliania NTP na pokazateli ekonomicheskogo razvitia', *Ekonomika I Matematicheskie Metody*, 2, pp. 252–266.
[24] Vincents, V. (1994), 'Imperfect competition in international trade: consequences for Eastern Europe', Paper presented to the conference on 'Alternative Ways of Economic Reform in Ukraine', Kiev, December 1–4.

CHANGING FACTORS OF TECHNOLOGY TRANSFER IN THE CZECH REPUBLIC

KAREL MÜLLER
Institute of Learning Foundations
Charles University, Prague

1. Introduction

This chapter analyses technology transfer within the framework and conditions of the Czech Republic. It focuses on the dynamic and structural aspects of regulatory and institutional issues, at present associated with essential change. This includes the transition from a centrally regulated to a competitive system , the redistribution of the economic power structure (itself linked to the privatisation of manufacturing industries and services), the redistribution of technological capacity from a hierarchical to an open (horizontal) pattern, and the internationalisation and globalisation of a national technological framework.

Analysis will include a careful assessment of institutional change, incorporating the link between the public and private sector and between economics and politics, and taking into account both their inhibiting and mobilising effects. This approach, by attempting to identify the formative capacities and agents in a situation of institutional change, is very much dependent on the conceptual framework and its methodological orientation—particularly in a situation of lagging empirical databases and insights. To help focus our analysis, attention will be concentrated upon technology transfer within the field of manufacturing industries only. The situation in agriculture, mining and the services sector will be discussed only indirectly.

2. Technology Transfer in the Conceptual Debate

This section describes the concepts of technology, and technology transfer, which may be relevant to the transition economies. Central and east European (CEE) countries face change to both their economic and political institutions and must mobilise their resources and personnel to attain structural changes in the economy and socio-political framework. Although current debate centres around the pressure of economic decline, increasing social tensions and the establishment of institutions for a market economy and political democracy, effective changes are mainly conditioned by the change in the knowledge and technology base.

At the outset, various strategic options for the interface of the technology base and institutional transformation in the CEE countries may be suggested:

J. Kirkland (ed.), Barriers to International Technology Transfer, 185–196.

- their national technology base could be used to further economic and democratic aims;
- it could become their constraining factor;
- it could shape social change within current socio-technical patterns by moving towards either liberal democratic or social democratic patterns [2]. Indeed, the issues of technology transfer will be shaped by the *institutional transformation of the national technology base*[1]—the changing interfaces between technology base institutions and economic and political institutions, and their socio-cultural restructuring.

Thus our understanding of the technological situation in the CEE countries gains little from the traditional neo-classical concept which treats technology as a 'black box', readily responding to the changes in factor productivity. Corrective considerations which might be associated with social/private returns, or the market/public assessment/failure in the promotion of technology, are not yet mature, as the market institutions and the efficient public regulatory bodies are only beginning to emerge. Better cognitive input can be expected from approaches which attempt to analyse the growth of technology in its socio-economic framework, its mutual interdependencies and in the entrenchment of technology capacity (and personnel) within economic and political institutions. The situation of the CEE countries demonstrates both the impact of available technology capacity on the shaping of market institutions, and the effective intervention of politics and concepts of reform, into the technological infrastructure. Here, results of evolutionary economics, and the study of science and technology, can have productive cognitive impact.

Several areas of social science would benefit from better insight into the role of institutions, fields of technology and their personnel. Technology growth is taking place in a competitive environment but the decision to innovate does not rest entirely on competitive advantage. Relationships between the various parts of the technology base are also of importance, using traditional concepts and policies. For example, in the neoclassical context, the assumptions of competition, rational decision making and perfect information describe the discrete elements of technology which can then be mobilised by a competitive environment and guided by market institutions and the reliable direction of private agents. Such an approach can guide the exploitation of existing technology patterns and promote the established agents of S&T, but it is of little use in understanding and assessing the implications of S&T, or in searching for ways of democratising the control of S&T institutions.

This knowledge is based on study of the changing links between technology and the economy, which provide the crucial momentum for economic growth. It is accompanied by a search for more effective forms of networking—by academics, industry and state [4], and by various regulatory factors involved in socio-cultural change. Technology is seen to be not only a source of economic growth or a political tool, but fundamental to the formation of cultural values. These values can only grow within a global context, providing reference is also made to local socio-cultural patterns. Democratisation of a technological culture is therefore closely associated with democratisation of a political culture [1][5].

Strategies for reform of the technology base in the CEE countries need to be studied carefully in the light of the trends outlined above, particularly with regard to their structural dependencies and mobilisation effects. The aims of institutional modernisation are constrained not so much by a lagging technological base, or scientific and engineering capabilities, as by economic potential and entrenched social attitudes. Technological capacity has been embedded in a hierarchical power structure; technology transfer, therefore, has been undertaken within a framework of specific government rules on funding and programming the key sources of technology growth—academic science, industrial research, investment, public distribution of goods and the import and export of goods. This method was known in the west in the 1950s and 1960s as the 'linear technology push model', though with the structural differences caused by the systematic interplay between economics and politics. In former socialist countries, however, such methods were associated with the deconstruction of the institutions of private property and the democratic political public. Formal and informal activity thus became divided, with motivation, competition and cooperation limited by formal, hierarchical patterns on the one hand and spontaneous localised effort existing unsupported on the other. In the 1980s such divergence in power distribution and coordination was apparent in all institutions and, at a local level, it produced alternative bursts of action and social tension. Those involved have formed the backbone of a new society, now undergoing political and economic change. Similar changes were being felt in the institutions of the technology base and, as their structure and potential are vital to a healthy competitive environment, they will be discussed more fully before an analysis is made of the present technological situation.

3. A Tentative National Model of Technology Transfer

Technology transfer in a centralised economy may be outlined with the help of a technology–push model. Here the market–pull, public control of technology and public adaptation to the challenges of technology are substituted for a 'rational', centrally-regulated, organisation with consumers, partners and citizens behaving in an expected and cooperative manner. Of course, such a 'grand technical system' was unrealistic in practice, although it had some utility during the postwar ideological mobilisation thanks to the state's redistribution of inherited economic resources. In practice, institutions adapted themselves to the changing balance between the formal central power and the rather uncoordinated actions of individuals.

In the 1970s and 1980s the dissipative pull was associated with and mobilised by the growth of new global technologies, industries and management practices, and influenced by the growing technological gap in favour of the advanced countries. Those involved in technology were searching for new options while the influence of state power was declining. A specific decentralisation of power developed in which the state lost its administrative influence to enterprises, regions and so on, which in turn were growing in power, competence and networking. We now examine the outcome of these developments within the national technology base.

The infrastructure of technology transfer has been shaped by traditional institutions and those involved in them. These include

- academic and industrial research bodies, based within educational and industrial institutions
- educational institutions with a tradition of engineering and training, developed

regional networks and links with industry
- production institutions, with a tradition of heavy and light industry and a network of industrial technology (raw materials, machinery, processing)[2]
- the state, its administrative capability and regulatory practice *vis-à-vis* self-awareness and strategies of subordinate institutions.

Compared with western democracies, the socialist model of technology transfer was in structural terms short of market control, an independent banking system and democratic political regulation. Technology transfer was inhibited by various sorts of communicative deficits and institutional barriers (between academia and industry, between producers and suppliers, between manufacturing branches, regions and so on). On the other hand, dissipative decentralising trends to a certain extent promoted technology transfer. They provided a background for informal horizontal networks, easing competition between institutions (in their struggle for state funding), and mobilising formal communication and some inter-institutional cooperation.

Beginning in the 1980s, specific examples of dynamising factors , conducive to technology transfer in Czechoslovakia, included the following:

i) Manufacturing. The growth of the funding power of enterprises,[3] their technology-based networks of mass production units, and the pressure of new technologies on those which were export-oriented. These units were increasingly in a position to extend their in-house research, promote outside R&D funding and develop the practices of contract research. Increasingly, forms of technological networking developed, showing a more obvious technological demand from industries. Such forms of technological coordination have been described elsewhere as a technological quasi-market [3].

ii) Research. The response of research institutions to the technological demand of industry has grown, as has alternative R&D funding in the institutes of the Academy of Sciences, HE establishments and mission-oriented industrial research institutes. Informal contacts between the research and production communities have been established.

iii) International aspects. The influence of technologists has grown, and international communication and contacts have begun. Although this was curtailed by politics, it developed informally; in particular it has influenced the capacities of the research institutes of the Academy of Sciences (with wider communications and exchange of scientists) and the export-oriented enterprises (with better contacts in world markets and the growth in licence purchasing).

These changes in the growth options of technology agents influenced their direction and patterns of communication and thus became embedded into the framework of the institutions, largely by differentiating between the research and professional communities. Similarly, growing differentiation in the practices of mass production units disclosed tensions in their production structure. These were caused by a lagging technological regime, low productivity and the relationship between production and environment, and production and consumption. The prevailing agents of technology transfer and the interfaces between them are set out in Figure 1.

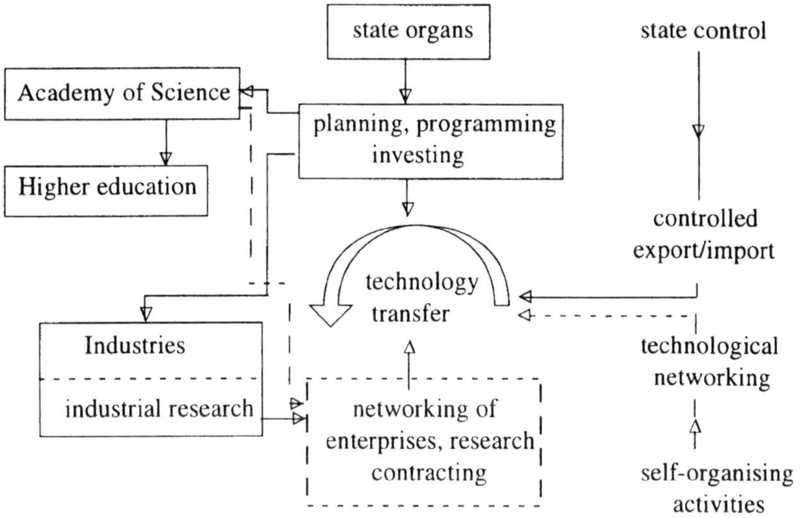

Figure 1. Model of technology transfer I (agents, links, impacts)
—— *strong impact* - - - - *weak impact*

4. Reformative efforts and their relevance to technology

Reformative effort in the Czech Republic is characterised by the radical and consistent transformation of the regulatory framework of economic and political systems. The economic approach has prevailed within the various concepts and agents of transformation, directing the transformation process towards the formation of market institutions and the standardised redistribution of economic power. Under the influence of the political pressure of democratisation, the deconstruction of former institutional and regulatory frameworks has begun. Next, and probably for good reason, the instruments of economic regulation have been applied combining microeconomic mobilisation with macroeconomic balance and restriction—massive liberalisation and privatisation of economics on the one hand and tight monetary and fiscal policy to balance the state budget, and anti-inflation policy, on the other.

The consistency of economic strategy has been based on a neoclassical and neoliberal framework and produces serious implications both for those spheres which are outside it, and the resistance of their agents who feel threatened by it, including the institutions of the technology base. In quantitative terms the effects of these developments have included the following:

- The academic and industrial research labour force has been reduced by half and its funding by more than two thirds (see tables 1 and 2).
- Educational establishments have grown in terms of student numbers and students redistributed in favour of social sciences and humanities (see table 3).
- Output in production has declined by 30 per cent but a slight recovery is beginning.

TABLE 1. Sources of R&D funding
(CR, GERD, millions of Kc, current prices)

Source	1990	1992	1993	1994
State	7,429	2,754	2,781	3,626
Enterprise	21,022	14,499	9,628	9,357

Source: Czech Statistical Office, Prague, 1995; foreign funds are included in both sources of funding.

TABLE 2. Distribution of R&D manpower by research sector
(CR, 1994/1992, thousands)

Ind/sec.	BES I	BES II	Govern.	AS	HE
1992	11.8	23.7	14.5	8.5	5.5
Share %	18.4	37.1	22.6	13.3	8.6
1994	7.4	15.6	5.0	6.1	3.4
Share %	19.7	41.3	13.6	16.3	9.1
1994/92 %	62.7	65.8	34.5	71.7	61.8

Source: Czech Statistical Office, Prague, 1995.
Note: The distribution by research sectors is adapted to the OECD pattern (and recalculated by the author since this classification is not yet fully applied in Czech statistical practice); BES I = independent industrial research institutes; BES II = in-house industrial research; AS = Academy of Sciences; HE = higher education.

TABLE 3. Structure of HE students by type of institution
(1991/1992, number of students, Czech Republic)

Type of school	Study	Per cent	Newly enrolled	Per cent
Total	93,384	100	21,737	100
Natural science	4,443	4.7	1,075	4.8
Technical science	33,633	36.0	8,086	37.5
Agricultural science	7,487	8.0	1,510	6.9
Medical science	10,695	11.4	1,509	6.9
Social science	36,343	38.9	9,088	41.7
of which				
Economics	11,804		2,408	
Law	3,326		1,075	
Pedagogics	16,398		4,091	
Arts	1,876	2.0	469	2.1

Source: Zeleny, P. (1993), *Structural changes of the Czech HE system towards its institutional diversification*, Institute of Sociology, Academy of Sciences, Prague.

TABLE 4. Changes in output and productivity in selected manufacturing branches (CR, per cent, 1989, 1993, 1994)

Period	1993/1989	1993/1989	1994/1993(a)	1994/1993(a)
Branch\in.	Output	Product.	Output	Product.
Industry total	58	81-82	99.9	105.5
of which				
Machinery, el.-tech. in.	45	66-67	100.8	109.4
Leather, text., cloth	49	70-71	101.1	106.3
Fer. metals	55	73-78	102.1	106.8
Wood ind.	59	61-63	98.7	98.3
Chem. & rubber	65	75-77	106.2	108.5
Glass/ceram.	67	80-81	103.2	106.5
Foodst. ind.	71	82-85	99.1	97 .8

Source: Czech Statistical Office, Pick 1994.
(a) January–September, 1994.

The benefits of this, however, have been uneven, and directed in favour of products with a comparative advantage from low-cost labour and established (medium- and low-level) industrial technologies and branches. Many high-tech branches of industry have collapsed and the productivity gap with the advanced countries has continued to grow (see table 4). On the other hand extensive change in the organisation and ownership of industry has occurred, with productivity starting to grow in some areas and in some new forms of ownership, such as reorganised large enterprises and joint ventures.

- State regulatory institutions in the field of S&T have been completely reconstructed. The relevant ministries were closed and activity brought into line with a level of state funding of 0.4 per cent of GDP. The Council for Science and Technology was established in an advisory role; the government's S&T policy was reshaped to support research and several grant agencies were established to distribute state R&D funds by means of public competition and grants. Public support schemes for technology agents, through SMEs, science parks and technology transfer, were established but are not influential. Vital frameworks for the growth of important technology agents are emerging through the formation of capital markets, and the independent banking system, but the impact of these on technological developments so far is limited. Counter pressures on banks have included the bad credits of larger industries (mostly with outdated technologies) which have restricted funds and therefore the capacity available to promote technological reconstruction.
- Enterprising activities are developing extensively within the rapidly growing private sector and the self-organising efforts of professional communities —examples include the growth of private consulting, the Technology Foundation, the Association for Technology Transfer and renewed engineering communities.

TABLE 5. Changes in number and size of enterprises

(by number of organisations and employees; 1990, 1992, 1993)

Size	Total	Less than 500	500-1000	1000-2500	Over 2500
1990(a)	1028	459	207	259	103
1990(b)	1272	106	140	391	645
1992(a)	2416	1765	299	245	107
1992(b)	1522	310	212	381	619
1993(a)	3109	2488	314	174	133
1993(b)	1337	365	219	245	508
1990/93(c)	259/420	99/315	65/47	65/37	30/21

Source: Statistical Yearbooks CSFR, C, 1991, 1992, 1994.

Notes: (a) number of organisations; (b) number of employees, (c) machinery industry, number of organisations.

The dynamising effects of the changing economic environment on the domestic technological base have contradictory implications. The outcome of these trends is not yet clear and it is thus difficult to identify qualitative changes to the domestic technological base. Fast (and mass) privatisation (using the coupon method) has caused dramatic growth in the business enterprise sector (BES), which has, however, by attracting technologists from the state sector, thus drained the capacities of existing enterprises and organisations. Public institutions taking care of the infrastructure of services have been particularly hit by these developments, together with a decrease in public funding.

The massive disintegration of existing enterprises has decreased their size (see table 5)and their accumulative potential, and has exposed their networks (including the technological ones) to the competitive environment. On the other hand, spin-off firms, or privatised expert services, could exploit the technology and expertise available by improving organisation patterns and sale practices, by higher motivation and by closer interaction with customers, but they have been suffering from the shortage of capital. The formation of the domestic technological market, and its liberalisation and globalisation, has promoted a competitive environment with the increase of agents at the supply side (albeit private and foreign agents), and the quality and choice of products. The demand side, however, has decreased dramatically, due to the lack of capital and investment. The consequent tension between the supply and demand sides of the domestic technological market thus undermines the formation of its institutional framework.

5. Factors Influencing Technology Transfer in Transition

The most prominent factor influencing technology transfer is clearly change in power distribution (from state to private/public sector) and the formation of the business enterprise sector. Private agents have been able to mobilise available technology resources by new interactions with the market place, or adopt new technologies with respect to available labour force capacity, technological and marketing networks and capital resources.

Global market pressures such as the collapse of the eastern-bloc market and growing competition in the domestic market has also played an important role.

The results of recent research indicate that firms with previous technological and marketing competence and networking, in both domestic and foreign markets,[4] have best survived economic recession. The extent to which these competences are capitalised upon, however, is conditioned by other factors which are not so far linked to a specific form of ownership but rather to the scale of the enterprise and its position in the international technology market. The former advantage lies with large enterprises with complex production programmes and a stable technological position (some of them still owned by the state), the latter by smaller and specialised (private) joint ventures which are attempting to improve their position in the technological chain. Both factors have played a decisive role in the survival of medium and low-level technology firms, since these have been able to redirect their exports to EC countries, mostly as suppliers of semi-finished products. The above mentioned cases of productivity growth also confirm the importance of the capital/technology interface, in the first case through better admission to domestic banking houses and in the second through the close relationship between foreign technology transfer and capital transfers from abroad.

The shortage of capital and low investment activity are critical to the reconstruction of a national technology base and that is why capital saving technologies and non-investment factors are of importance. Their use is related to the disintegration process of Czech industry: the medium-sized and smaller companies are usually more flexible in restructuring their organisations to increase efficiency and effectiveness in relation to new market situations, and also in raising the technological level of their products. However, the capital saving technology efforts seen to date reflect better utilisation or improvement of existing technology, rather than the type of reconstruction which is crucial for economic survival in the new environment.

While the role of domestic experts and expertise is growing dramatically in the formation of the business enterprise sector, their capacity within public technological institutions (for example, research and education institutions, information services, public support schemes and other public services directed at the long-term growth of human capital) is declining. It is not clear so far to what extent such development will result in the reinstitutionalisation of public services.

The boundary between the public and private sectors is largely shaped by the aim of fiscal policy, which has deliberately maintained a hard and fast line between the two sectors. This approach is required for efficient monetary policy and the deconstruction of the prevailing state-owned and excessive welfare institutions, but it does not help the search for new institutional settings. The growth of intermediaries between public and private sectors is thus of central concern to education and research institutions. Here also contradictory developments may be observed. One example is the forthcoming regionalisation of higher education, which can be a productive intermediary between the education and production sectors locally. However, the constraints of the legal, administrative and funding systems do not allow such pockets of change to grow. Similar developments have affected the academy–industry relationship, where the capacities of research institutions to adapt to the changing situation are constrained by the simplified route of the transformation of research institutions.[5] Industrial research has been transferred to the business enterprise

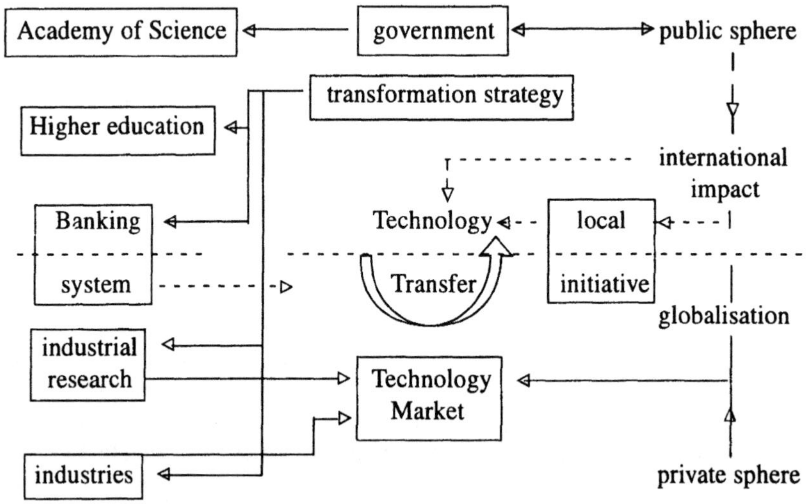

Figure 2. Model of technology transfer II (agents, links, impacts)
—— *strong impact* - - - - *weak impact*

sector, and exposed to severe selective pressure from the technological market, while strategic research capacities are left to the care of the fragmented and competing interests of the research communities. Figure 2 gives an overview of the agents of technology transfer and the interfaces between them, which are emerging alongside the transformation process.

6. Discussion

From the above it will be seen that attempts to improve technology transfer face several contradictory pressures. On the one hand, the radical and liberal approach of transformation strategy could mobilise better use of existing technologies or the search for advanced ones. On the other hand, traditional technologies and their industrial networks have used the liberal environment to strengthen their position using various sources of comparative advantage. Whilst the technological base could be mobilised in favour of short-term economic aims, the impact of the new economic environment on the growth (and restructure) of the technological base has been so far negligible.

The impact of these changes on the national technological base also reveals contrasts. To a large extent the massive redistribution of technology capacity in favour of market forces is a positive development, since it helps correct distortions created by the regulatory pattern of the S&T push model. But such developments have also created new tensions and imbalances, particularly in the relationship between public and private institutions, as the former adapt to their weakened public authority and emerging competition in the political system.

These developments must also be seen in the context of attempts to pursue a consistent economic and fiscal policy. The effect of such policies on providing and focusing technology transfer has, however, been limited. Whilst they have influenced the course of deinstitutionalisation, they have been unable to identify and promote niches of reinstitutionalisation.

In contrast, the effect of bottom-up activity is growing, but with conflicting effects. These can be seen not only in the public/private interface but also in other structural dimensions, including the interface of technological dependencies and the emerging competitive technological markets and democratic control of the economic sector. The newly emerging structural imbalances are undermining the formation of market institutions (with various forms of monopolies and unfair competition) and are also deconstructing the informal networks and the options for positive regulatory and self-regulatory patterns of action.

These competitive and destructive elements are necessary symptoms of institutional change, and can be justified in the context of the CEE countries. However, modern technology and its democratic regulation are based on the cumulative and socially entrenched networks which need a great deal of reflexive activity in order to balance their proportions—a space for sound technology growth and diffusion and the power of its economic and social appropriation.

Notes

1 The notion 'technology base' is used in a pragmatic way in order to enable the analysis of the relevant institutions and their interfaces; the make-up of the technology base institutions is discussed in section 3, suggesting a model of technology transfer.

2 The distribution of the manufacturing industries in the Czech lands among the primary, secondary and tertiary sectors was influenced by strong technological relationships which were related to domestic raw materials (agriculture, kaolin, glass sands and coal), their processing (food industry, chemical industry) and mechanisation (machinery and electrotechnical production and its feedback to metallurgy). Also some weak links were established depending on the availability of raw materials (iron ore, cotton for the textile industry, leather for the shoe industry) and fluctuations of the market (military production).

3 The self-funding ability of mass production units was, however, limited to current expenditure (including R&D, which became absorbed into the 'Fund of Technological Development') and smaller capital outlay. Essential capital outlay was centralised and redistributed by the state.

4 The EC project 'Technological modernisation and the combinates', coordinated by CERNA, Paris; its aim is to study the relationship of the disintegration and reorganisation of large enterprises to their technological reconstruction.

5 The analysis of this issue has been studied with the help of the EC project focusing on the academy–industry relationship in selected EEC countries, and coordinated by H. Etzkowitz, A. Webster and K. Balazs.

References

[1] Bijker Wiebe, E. (1995), 'Democratization of technology, who are experts?', paper presented at seminar '*Expertenkultur und Demokratie*', Aachen, 23 February.

[2] Machonin, P. (1993), *General Approaches to the Problem of Post-Communist Social Transformation*, Prague, Institute of Sociology.

[3] Mueller, K. (1995), *Changes on the 'Borderlines' between Research and Industry following the Economic Transformation in the Czech Republic*, Social Studies of Science.

[4] OECD (1992), *Technology and Economy, The Key Relationship*, Paris.

[5] Schwarz, M. (1993), 'The technological culture: challenges for technology assessment and policy', *Science and Public Policy*, 20, December, pp. 1–8.

ACADEMIC ENTREPRENEURSHIP IN HIGH-TECHNOLOGY FIRMS IN BULGARIA

KIRIL TODOROV
Department of Industrial Business and Entrepreneurship
University of National and World Economy, Students Town, 1100 Sofia
Bulgaria

1. Introduction

The formation of small and medium enterprises has been one of the most striking indications of entrepreneurial activity in the transitional economies of central and eastern Europe. In Bulgaria, which has a population of under 9 million, approximately 360,000 such enterprises have been established. The reasons for this development, at a time when government regulation and institutional support for SMEs are at an early stage of development, is an important topic for research.[1]

As in most eastern European states, the majority of the new SMEs operate in the field of trade, services and transportation. The average share of industrial SMEs is under 10 per cent. One category of enterprise not fully covered by the current literature is high-technology SMEs, generated to a large extent by so-called 'academic entrepreneurs'. From the available literature on academic entrepreneurship, it is clear that these individuals and companies place emphasis on the transfer of scientific knowledge from the academic world to the real economy and, in particular, industry. Other research shows that the majority of such firms remains small, so that their study is complementary to that of SMEs in the economy as a whole. Finally, it is clear from past work that academic entrepreneurs display extremely high motivation.

This chapter will describe the current state of academic entrepreneurship in Bulgaria, and the environment in which it exists. It will also draw out implications for the channels of technological and entrepreneurial know-how transfer. The analysis will be backed up by presentation of a typical case study.

2. The Environment for Academic Entrepreneurship

As with other areas of Bulgarian life, the development of academic entrepreneurship can be divided into two main stages—before and after 10 November 1989, the day on which the transition towards a market economy can be said to have started.

Before 1989, the environment was characterised by strong centralisation of public economic life, the lack of effective market mechanisms and any meaningful conditions

J. Kirkland (ed.), Barriers to International Technology Transfer, 197–205.

for entrepreneurial activity. These factors also applied to academic institutions and specialised branches of scientific and science production institutes, which were the main generators of technological know-how. Academic know-how, at least on fundamental research, was primarily generated in laboratories and institutes of traditional high schools, in particular Sofia, Plovdiv and to some extent Schoumen Universities. More practically orientated innovations, including the development of new or updated technologies, were prioritised by technical high schools such as Sofia Technical University, Varna and Gabrovo Technical Universities. Specialised institutes of different ministries and larger business units—including associations, unions and plants—were most frequently used as a channel of transferring academic know-how into practice.

The key problems of academic entrepreneurship under this system at macro and micro level can be set out as follows:

At macro level:	*At micro level:*
- lack of a national system for generation, transfer and diffusion of know-how, including academic	- low level of technological knowledge, dependent largely on the achievements of Soviet and other COMECOM member states
- insufficient regulation to ensure protection of intellectual property (an issue to which academic innovators and entrepreneurs are very sensitive)	- absence of entrepreneurial skills necessary for building up effective market behaviour, formulation and realisation of competitive strategies etc
- lack of private property and real market conditions, contributing to development of entrepreneurial activity	- low motivation of potential academic innovators and entrepreneurs due to the inability to gain real recognition at the market and the unprotected copyright
- inadequate conditions for innovations and entrepreneurship in the universities themselves, including availability of highly effective special equipment and access to developed know-how of industrial countries	

The changes instigated from 1989 radically affected the status and behaviour of academic institutions, and indirectly the academic entrepreneurs within them. The positive and negative characteristics of these changes are summarised on the following page, from which it can be seen that the effects of change have by no means all been positive.

One consequence which has become increasingly evident during the past year is the exposure of technical high schools, which are experiencing more difficulty in recruiting students, and in retaining lecturers and researchers with entrepreneurial tendencies, while leaving for commercial jobs in search of larger incomes.

One interesting development concerns the behaviour of high-level students in trying to seek careers without relying exclusively on support from their high schools. One alternative increasingly utilised by such students is to enter the optional faculties of technical or economic high schools, where they can acquire specific (and lacking) expertise in economics, management and increasingly specialised entrepreneurial skills. There is some evidence that students/innovators from technical high schools and business students from economic high schools, who have become acquainted through courses in such optional

At macro level:

positive changes	*negative changes*
- General liberalisation of economic & social activity in society	- Restrictive economic policy (high interest rates, inflation ratio, taxes)
- Adoption of new copyright & patent law	- Science and Higher Education law not passed until recently
- Introduction of high school's academic autonomy law	- Rapid restriction of the high schools' budget connected with overall economic difficulties
- Creation of additional possibilities for high schools by introduction of paid form of education (besides state ordered education)	- Exodus of young researchers, lecturers and doctoral students, esp. from technical high schools (typical representatives of academic entrepreneurship) observed as consequence of above changes.

At micro level:

positive changes	*negative changes*
- Repeatedly enlarged motivation for achieving results among academic community	- Insufficient entrepreneurial experience - especially among technical high schools
- Improved communication capabilities (mass-media, satellite connections, e-mail, Internet, etc.)	- Lack of sources of finance (especially concerning risky projects)
- Greater possibilities for individual progress - participation in international projects, including PHARE programme	- Misunderstanding by senior lecturers and researchers of 'hotheads'
	- Bureaucracy at university level during introduction of innovations and their transfer into commercial products (and consequent distribution of earnings)

faculties, are now beginning to cooperate in the establishment of private firms.

Finally, it is clear that one of the major obstacles to academic entrepreneurship has hitherto been a prevailing reticence within the framework of the former COMECOM. More recently, it has become clear that the specific features of Bulgaria, notably its small size, scarcity of raw materials and strategic geographic location, are serving to impose an open character on the economy. Against this background, a positive attitude towards the achievements of innovators and academic entrepreneurs from developed industrial countries will be critical to further progress.

3. Transfer of Entrepreneurial Know-how and its Impact on Academic Entrepreneurship

Effective mechanisms for transfer of entrepreneurial know-how, including transfer from other countries, is essential to the development of academic entrepreneurship. In this context, it is essential to define key terms and hypotheses, which can then be used as working definitions.

There is no single definition of knowledge transfer. We can, however, point to the work of Rogers [3], who described technology transfer as ‘a process thanks to which an innovation reaches members of social systems in time through communication channels’. Rogers argues that the choice of transfer model is influenced by several factors, including communication channels, time and human factors, and the prevailing social system. The existence of an innovation culture is critical to effective transfer. Innovation takes place everywhere in the modern economy, and can include the creation of something new by combining existing elements of knowledge in a new way [2]. This view is critical to our current analysis for several reasons. Firstly, it describes innovation as a cumulative process. Secondly, it emphasises that creativity can involve the combining of existing knowledge as well as the creation of new knowledge. Third, it emphasises that the cumulative process includes the adaptation and spread of knowledge as well as its creation. Thus the mode of analysis described takes us closer to the nature of entrepreneurial activity, which also draws on more modern approaches to previous innovations. Thus we can interpret knowledge transfer as embracing both the generation of new and innovative practical components, and the practical measures required to adapt and to enter new markets.

The potential of key individuals to absorb knowledge is fundamental to the transfer process. This depends on many internal and external factors. External factors include levels of education, history and culture (which itself may draw on traditions, religion and psychology). Internal factors might include company type, past activities and quality and motivation of staff, especially at managerial level. The quality of staff, and especially of its entrepreneurs/managers is of particular importance for the small company.

In view of the importance which must be attached to international technology transfer in the Bulgarian context, the extent to which know-how can be adapted between spheres and countries is vital. In this context, Johnson and Lundvall [2] discuss whether transfer from one cultural environment to another is possible, and if so whether this can help to increase know-how effectiveness. They point out that Denmark and Sweden are countries with similar histories and cultures, but very different innovation and institutional structures. Development of their hypothesis would be useful for advancing the theory of know-how transfer in an international context. This could greatly assist in the development of models for transfer between countries which are similar in terms of territory and population, but different in culture and psychology, thus helping to overcome cultural and psychological barriers which currently exist.

The channels available for transfer to take place are critical to its success. These can be formal and informal, direct and indirect, or a combination. Practice shows that it is much better to transfer and adapt currently available know-how (although this may not be at the leading edge) than to transfer such advanced know-how incorrectly. It is clear that formal channels are of the greatest importance since these are the mechanisms for transferring standardised knowledge. We can study such formal channels at two levels: the macro level, represented by national systems and programmes transfer (carried out by institutions such as ministries, committees and so on) and the micro level, which is represented by universities, colleges, business schools, consultant companies, founda-

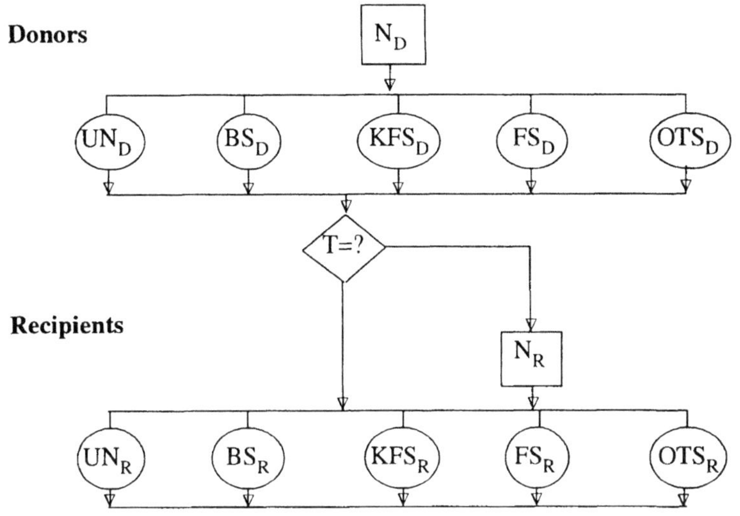

UN = universities and similar educational institutions
BS = business schools (training institutions)
KFS = consulting firms
FS = real business firms
OTS = others (associations of SMEs, foundations, etc.)
N = national institutions and programmes
D, R = indexes
T=? = transfer of entrepreneurial know-how: from where, what, how, who?

Figure 1. The process of transfer

tions and so on. Each unit in the process can be characterised as a donor or recipient according to its position. The general transfer process can thus be illustrated as in Figure 1. It is clear the transfer can pass indirectly through intermediate units or involve a direct relationship between donor and recipient, for example a university and company.

The agents of transfer (such as lecturers, researchers, consultants, entrepreneurs and managers) play a critical role in the case of entrepreneurial know-how because of the influence of personal characteristics. Again, the role can be formal or informal. The informal role is becoming more important in the transitional period, as it becomes clear that institutionalisation alone is not sufficient. Strong relations which develop between donors and recipients can later on be institutionalised, thus becoming formal channels for transfer.

Where a great number of transfer channels and agents exists, problems can develop in their coordination, standardisation of knowledge transferred and its spread amongst recipients. These difficulties are reinforced by entrepreneurial heterogeneity and the large number of potential recipients.

From existing theory, and our own studies, we can conclude that Bulgaria lacks the theoretical, systematic and methodological background for entrepreneurial knowledge transfer. It is therefore important to try and adapt present models and transfer practice

TABLE 1. Relations in the process of transfer

Transfer channels	Relations 'donors'-'recipients'	Content and form of know-how transferred	Evaluation of effectiveness
N	ND-NR	Conceptual and system knowledge; work and carry out national institutional branch and regional programmes on SB	L
UN	UND-UNR	Training young well-educated entrepreneurs training programmes and projects, training of trainers	A
	UND-BSR	Training of trainers - directly (when participating in training), or indirectly (programmes)	L
BS	BSD-BSR	Methodical support for training in SB and entrepreneurship training of trainers	H
	BSD-FSR	Practical training of entrepreneurs (beginners and non-beginners); training in donor's country	A
	BSD-KFSR	Training and consultancy advice; training consultants	A
KFS	KFSD-KFSR	Direct transfer of consultant know-how having in mind local specificity; joint ventures	H
	KFSD-FSR	Direct consultant help and help in search of new markets	A
FS	FSD-FSR	Know-how transfer in different partnership forms: licence, franchise, joint ventures	H
OTS	OTSD-OTSR	Branch and regional support to entrepreneurs; finding the right partner and participation in three-part system	A

H = high level of interaction and effectiveness
A = average level of interaction and effectiveness
L = low level of interaction and effectiveness

from western Europe (for example, US to western Europe, Japan to US—see [5]) in relation to product and technological aspects. In the area of personal and psychological aspects, characteristics and behaviour, the appropriate reaction depends to a greater extent on the local situation and here we must rely on the experience, intuition and expertise of transfer agents.

Figure 1, which has been compiled on the basis of past experience and practice, helps us to obtain an insight into the nature, form and type of relations in the process of transfer. It will be clear that universities have an unenviable role, as demonstrated by the results achieved by academic entrepreneurship until recently [4].

4. The Academic Entrepreneur in Practice: the case of TRIBO[2]

The difficulties faced by academic entrepreneurs in practice can be illustrated by the case of Boris Rizov. Mr Rizov, a graduate of Sofia Technical University, was working as a probationer in the Institute of Metal Ceramics when he developed a new technology of friction discs for heavy freight vehicles and road construction machines. The solution involved a cover made with a new alloy, utilising local raw materials. Based on usage of Bulgarian equipment and instruments, the technology proved cheaper than western alternatives and more flexible with regard to production quantities.

Despite his experience of bureaucracy and the indifference on the part of managers of state-owned enterprises, Mr Rizov remained convinced of the technical and economic value of the invention. He therefore decided to bring it onto the market personally. In this context he was able to use the legal opportunties for establishing private production which existed in decree N 56 adopted in 1989, and he registered sole proprietorship in November of that year. In addition to his own enterprising entrepreneurial spirit, he was supported by his wife, an engineer by education and small business manager by specialisation. The involvement of further partners was deliberately avoided to prevent future uncertainty in the management and development of the company.

The establishment of the company on the precise date recognised as the beginning of the transition towards a Bulgarian market economy is more than just coincidental. One effect of political change was an interruption in the import of brake discs from the major previous supplier, in Poland. This did not, however, affect domestic demand for the products.

This favourable external environment clearly presented an opportunity for the new company. But with self-confidence and enthusiasm alone, the proprietor was able to gain an unguaranteed bank loan and, prior to receiving the order for discs from his largest potential client, he began to order production equipment. After successful exploitation tests, but before the establishment of a production base, he signed a contract for his products with the Tersnab company—the biggest Bulgarian supplier of spare parts.

Contrary to expectations, the initial financing was followed by tense months of repairs, building activities, delays in equipment delivery and development of products to reach the necessary quality standard. In 1991, soon after the company had begun trading, it was exposed to new threats and opportunities. Decline in the economy led to a reduction in demand, and deferment of payments. But at the same time niches occurred in the trade involving other types of spare parts based on similar technology. The company made a strategic decision to diversify its range of high-tech products. Whilst production is focused on company know-how, preparatory and labour-consuming operations were subcontracted. Thus the strategy for company growth was broadened to involve wider utilisation of subcontractors' capacities, by controlling the overall output of the process.

The liquidity problems described highlight the dangers of dependence on a single product line. For TRIBO, the decision to enter new business unrelated to their main field of activity was a forced diversification caused by negative cash inflows and restrictive credit policy. The liquidity problem was further resolved by the purchase of a food store in an attractive location, which raised the credibility for banks and reduced overall financial risk.

Further potential changes involved diversification into a wider range of spare parts for road/construction vehicles. In this area, the company is pursuing joint products with producers of military vehicles, from which it would gain from association with a high-profile supplier to the defence industry—seen as a symbol of high quality and stability. The achievement of these objectives would contribute towards a planned entrance into the external market.

In summary, the company began with three workers and had as its initial objective recognition in the Bulgarian market. More recently, this has grown to twenty-one employees in a considerably diversified business. It is looking to the future with confidence.

The case is also an illustration of academic entrepreneurship more generally. It demonstrates the importance to lecturers of assigning relevant scientific tasks to high quality students. It also shows how a raw invention with potential, based on local raw materials and competitive in cost terms, can be transferred into real practice.

It also indicates the changing nature of obstacles faced by Bulgarian entrepreneurs. Before 1989, these were largely connected with bureaucratic difficulties in state institutions and problems with the protection of intellectual property. Since 1989, more prominent difficulties have involved high inflation, expensive credits and lack of effective government support for inventors and entrepreneurs. For this reason the company has been forced to implement a strategy of diversification, involving purchases unrelated to its core business. It was only by this means that the proprietor was able to continue his innovative work in order to reach his ambition of becoming a supplier to the military industry.

In many ways, the case illustrates positive changes in the Bulgarian economy, including much greater freedom and opportunity for self-fulfilment involving the achievement of results. However, it also demonstrates the need for flexibility, and dangers of over-dependence on specific products or the domestic market. It must also be said that the individuals concerned were able to draw on an exceptional range of personal and business skills. It was the combination between these which was the foundation of the company's success.

By a policy of practical adjustment to changing circumstances, TRIBO now has an opportunity for growth and success. Without the range of skills available, and real support from government institutions, however, companies of this kind will continue to have difficulties due to the inhospitable business environment which still prevails.

5. Conclusion

The above study highlights key problems encountered by Bulgarian entrepreneurs, particularly those from an academic background, in pursuing the commercialisation of their inventions. Particular emphasis has been placed on high-technology firms. It is concluded that, whilst the enthusiasm and motivation of some industrial academic entrepreneurs can produce positive results, most will be doomed to failure without external help. Greater support is therefore needed both on a specific and general level, in identifying industrial entrepreneurs with the potential to succeed, and ensuring that the structure and culture of universities and government institutions is more responsive to their needs.

Notes

1 The problems of 'entrepreneurship', typical of some of the large enterprises, are not considered here.

2 The short version of the case is developed for the needs of the teaching process in entrepreneurship by Assoc. Prof. Kiril Todorov, PhD, Milen Baltov, Kostadin Kolarov and Todor Angelov.

References

[1] Jones-Evans, D. (1994), 'Proposed project into academic entrepreneurship in Central and Eastern Europe', ACE Programme.

[2] Johnson, B. and Lundvall, B. (1992), *Catching Up and Institutional Learning under Post Socialism,* Institute of Production, Aalborg University.

[3] Rogers, E. M. (1983), *Diffusion of Innovation*, Free Press, New York.

[4] Todorov, K. (1993), 'In search of Golden Fleece: or something more about transferring entrepreneurial know-how in the period of transformation', paper presented at the IntEnt'93, Vienna, 5–7 July.

[5] Whitley, R. (1991), 'The social construction of business systems in East Asia', *Organization Studies*, 12, pp. 1–29.

CONCLUSIONS

ADOPTION AND ADAPTATION OF TECHNOLOGY TRANSFER MECHANISMS BETWEEN NATIONS

L.E. PARKER
National Science Foundation
4201 Wilson Blvd., Arlington, VA 22230 USA

For decades, individuals and groups seeking to foster economic development through technology transfer from higher education institutions (HEIs) to industry have looked to successful examples of industry–university interaction as models to emulate. In many cases, these models were developed in North America and western Europe. In the majority of cases, replicas outside the originating country have not been exact copies of the originals. This chapter examines the phenomenon of customisation when modules are transposed from one country to another. The first section presents examples of model replication which illustrate situation-specific tailoring to meet local conditions and needs. In the second section, generic conditions that hinder effective industry–university relations (IUR) are explored as additional factors that can be relevant to the transplantation process. The final sections consider implications for future adoption and adaptation of IUR models. Unless noted otherwise, the term 'industry' refers to private sector firms.

1. Models in Translation

Western developed countries have evolved a wide variety of mechanisms and approaches, or models, for engaging in IUR. There are many approaches available for developing and other countries in transition to adopt, ranging from modest, informal person-to-person interactions to regional economic development schemes to complex national programmes. In the process of replicating an existing model, there are nearly always situations in the new setting which differ from those at the original site. Experience shows that the process of adopting models has also been a process of translating models to fit the new environment through adaptation of specific elements of the approach.

This section presents four models ranging in size and complexity that have been adopted outside the originating country. Where lessons learned about operating a specific approach are available, they are included as well.

1.1. UNIVERSITY-BASED RESEARCH CENTRES

Research centres and institutes have existed at US universities for over a century, and even longer in western Europe. Many have at least two sources of support, for example,

J. Kirkland (ed.), Barriers to International Technology Transfer, 209–226.

industry, government agencies and private foundations. During the preceding decade, there was much experimentation with the centres mode of academic research in a number of countries. In general, centres are intended to conduct research whose complexity, scale, and instrumentation requirements are beyond the capability of individual investigators or small research groups. In many cases, centres are expected to develop long-term relationships with industry and conduct interdisciplinary research that is relevant to industry. Some government-supported centres programmes expected centres to conduct very applied research jointly with companies.

University-based centres were an important component in US Federal and state-level strategies developed during the 1980s to enhance university–industry cooperation. The focus of state-supported centres programmes ranged from fundamental research to advanced technology development. In most cases, support came from a variety of sources, for example, universities, state government, and industry. While state-level programmes tended to focus mostly on technology development, those funded by the Federal Government placed greater emphasis on research of relevance to industry.

1.1.1. *US: Engineering Research Centres*

During the 1980s, the US National Science Foundation established a number of university-based centre programmes. One of these programmes, the Engineering Research Centres (ERC) programme, supports interdisciplinary research that extends knowledge in a variety of technology areas and is relevant to industry. The main programme goal is to bring engineering and scientific disciplines together to address research issues crucial to the next generation of technological advances using an engineering systems perspective. In addition, each centre must have active support from and participation by and long-term associations with industry and other user organisations [38].

In the 1993–4 award year, over 600 companies had partnerships with the eighteen centres. At present there are 21 centres from one to eleven years in age. Beyond the research and industrial involvement goals of the programme, there is a strong emphasis on involving undergraduates in centre activities to provide them with experience working in interdisciplinary teams, often with industrial researchers at both the host university and in industrial laboratories. Both the undergraduate and graduate students who are involved in ERC research are expected to constitute a 'new breed of engineers' who are prepared for working in industry in problem-oriented teams [39]. Two formal evaluations are now underway to examine how well the programme is performing in relation to its goals.

This model has been replicated in a number of countries.The following is an example.

1.1.2. *South Korea: Engineering Research Centres*

In 1989, South Korea established the Engineering Research Centres (ERCs) programme for development of new technology relevant to industrial applications. Funded by the Korean Science and Engineering Foundation (KOSEF), these centres are similar to their US namesakes [22, 24]. The objectives of the programme are to: 'improve capability for self-support in technology development by organizing scientific manpower in specific research areas at universities, by activating basic research and by fostering talent; . . .

[and] to reinforce international competitiveness through inductrial [sic] innovation to fulfill social and economical needs and to support national security in collaboration with those agencies overseeing public welfare, balanced development and national defense' [23, p. 3].

During the initial years, a centre is expected to increase in research quality and international visibility; disseminate information via seminars, workshops, intensive training, talks at conferences, and publications; and provide continuing education for and collaboration with staff in industry and government research institutes [24]. As of 1994, there were 21 ERCs across the country [23].

While research is the primary focus, education is a close second. The centres are intended to provide excellent preparation for students who, in ten years, will lead the country's development. In order to improve the quality of research and professors at the centres, KOSEF intends to entice progressively more Korean students who would have studied abroad to study at domestic universities with ERCs instead. Programme designers also intended for the centres to play a significant role in reversing the country's earlier brain-drain problem by conducting research that is sufficiently interesting to convince researchers who stayed abroad after schooling to return home [25].

This programme is noteworthy. While in its design it is similar to the US version, it is tailored to meet the needs of South Korea. In many ways, it is quite different from the original, particularly in terms of expectations about the Korean centres' role in technology development and national defence. The Korean ERCs are clearly elements of a national strategy for strengthening research, education, and technology development capabilities. Goals are clear and incentives for participation are built into the programme design.

1.1.3. *Lessons Learned*

There are a number of factors that can affect how successful an individual centre involved in IUR is going to be. Firstly, the university's culture is important, specifically, whether it can be characterised as 'entrepreneurial' [48, p. 18]. In the absence of this feature, it is difficult to encourage faculty members, especially younger ones, to be involved in a centre because of fears that centre-based research could reduce their chances for tenure and promotion. When a centre's programme is intended to have an effect on the culture of academic research, as the two mentioned above clearly are, it is critical that faculty in relevant departments feel free to be associated with centres. The more interdisciplinary the centre's technology area, the more important it is that institutional policies and values provide incentives for both involvement in centre research and working in interdisciplinary research teams that focus on topics of relevance to industry.

The vision of the centre director has significant impact on a centre's success. The director must have a clearly articulated sense of where the centre should be going in the future. The vision provides the intellectual direction for the centre and is the glue that binds the centre's research teams and projects together into an integrated whole that is more than the sum of its component parts. A strategic plan that emanates from the director's vision provides the road map for the centre. Without a strong commitment by the director to the vision, a centre can lose its synergy and integration, thereby degenerating into a collection of discrete projects. Once this happens, a centre is a centre in name only.

In strong industry–university research centres, the role of industry is multifaceted. Beyond providing cash and in-kind donations to centre operations, industry personnel can be involved in such things as: (1) providing advice to the director about the centre's generic research projects agenda, (2) developing shared testing facilities, and (3) personnel exchanges in which industrial researchers work with the centre's research teams and centre faculty and students spend time in company research facilities. Personnel exchanges are crucial to a centre's vitality. While financial support for both generic and proprietary projects is important, it is certainly not sufficient for true long-term, collaborative relationships between a centre and participating companies.

Many who study IUR use the terms 'cooperative' and 'collaborative' interchangeably. However, in describing characteristics of a strong versus mediocre centre, it is useful to think of these terms as expressing qualitatively different types of relationships. While provision of financial and in-kind support can be components of both, a strong centre whose relationship with participating companies is *collaborative* has significant movement of people back and forth between the centre and industrial laboratories to work together on joint projects and share equipment and facilities. On the other hand, *cooperative* relationships would be those involving mostly arms-length interactions with little or no joint research activities or exchanges of personnel [43]. Collaborative relationships are built on a foundation of many individual relationships between centre researchers and industrial researchers. Without strong individualities across the sectors, strong collaborations are unlikely to form [32, 37, 41, 53]. Ultimately, the key to long-term relationships is the continuing relevance of a centre's core activities to participating companies.

1.2. SCIENCE OR TECHNOLOGY PARKS

During the 1980s, one of the most popular approaches to IUR was the science park, also known as technopark or technology park. A few countries have had several decades of experience with this approach and a number of parks are at least a decade old [5, 17, 53]. Because of the variation in terminology, *science or technology park* will refer to an activity, programme, or facility in which high-technology firms establish operations on a large parcel of land on, or adjacent to, a university or university-affiliated research institute for the sake of collaborating with the host institution, as well as with other participating firms.

1.2.1. *US: Research Triangle Park*

Among the oldest science parks in developed countries is Research Triangle Park (RTP) in North Carolina. Established in the late 1950s, the Park exists within a triangle formed by the University of North Carolina at Chapel Hill, North Carolina State University at Raleigh, and Duke University in Durham. When the Park began, the Triangle region had been hard hit by the declines in the textile industry. Initiative for development of RTP came from the governor of the state of North Carolina. From the start, development of the Park has been managed, rather than spontaneous. Initial growth was slow, possibly because, unlike other early parks, it did not contain a premier research university when the park was established. RTP began to make significant progress in 1965 as a result of the announcement by IBM of its intention to establish a major Research and Development R&D facility at the Park. This led to the creation of 9,000 jobs and encouraged other large technology-intensive firms to establish R&D operations at RTP. IBM's initial move amounted to a vote of confidence that was vital to the success and longevity of the Park [34]. Since then, the number of companies inside—and increasingly just outside the Park's boundaries—have burgeoned. Requirements for firms inside the Park to engage in a specified amount of research led to companies not intending to conduct the requisite amount of research locating immediately outside the Park's grounds. The Research Triangle area has become a strong magnet for companies and individuals interested in taking advantage of all that the area now offers, especially for high-technology pursuits.

This Park is different from others in the US. Firstly, as mentioned above, the associated universities were not top ranked research universities at the time that RTP began. Secondly, the physical climate in the Triangle area is less attractive than that of other parks. Thirdly, as previously mentioned, the Park's development has always been managed [34]. Finally, initiative to establish it came from the state [40]. The first three are noteworthy because they go against traditional thinking about the characteristics that are important for attracting companies to join a Park. Nonetheless, there are two significant similarities with other parks. Firstly, initiation of the parks and early support came from strong, determined individuals. Secondly, they might have been deemed failures had the parks been evaluated too early.

1.2.2. *Taiwan: Hsinchu Science-Based Industrial Park*

Taiwan, like other Newly Industrial Countries (NICs), has few large domestic companies. Small- and medium-sized enterprises (SMEs) have played a key role in the acceleration of Taiwanese economic growth. Expansion of the manufacturing sector has been coupled with strong governmental emphasis on increasing the educational standard of the workforce. Rapid improvement in the quality of the workforce has enabled a transformation in the industrial structure from labour-intensive to technology- and capital-intensive [11].

As part of its industrial restructuring activities, Taiwan established the Hsinchu Science-Based Industrial Park (SBIP) in 1980. By 1989, the government had spent approximately US$320 million for land purchases, construction, and personnel. Three key components of the park are the Industrial Technology Research Institute (ITRI), National Chiao Tung University, and the National Tsing Hua University [47].

ITRI, a non-profit corporation, was established 'to act as a bridge between academic institutions and industries, [and] actively engage in the research and development of industrial technologies' [47, p. 4]. Organisationally, ITRI consists of seven laboratories dealing with a broad range of science and engineering areas and provides such services for industry as information dissemination and holding seminars to spread technical information [11].

By 1989, 105 companies had been approved as participants, of which 98 were operating and 79 were making products. More than half of the approved companies were new start-ups, and most of the operating companies employed 100 or fewer workers. More than 16,000 people were employed by Park enterprises, with approximately 40 per cent of them having at least a baccalaureate degree [17,18].

1.2.3 *Lessons Learned*

Regardless of how success is defined, the success of science or technology parks is mixed. Belgium provides an example. Ten science parks have been established since 1972. Some of the older ones are fully occupied by private sector firms. From this perspective, the parks are fairly successful. However, the extent of IUR presents a different picture. A study of collaboration showed that only 9 per cent of researchers working on industrial projects at universities housing science parks acknowledged any interaction with firms connected with the park. Conversely, of the firms associated with the park that participated in the study, more than half had R&D activities, but 32 per cent had no link with any university. What interaction took place was usually informal. The vast majority of the firms with connections with the host university also had connections with other universities in Belgium, other European countries, and the US [53].

Two principal formal means of establishing links in the science park context are academic start-ups, also called 'spin-offs', and what is known as 'tapping in'. Academic start-ups consist of firms created by faculty members who take their ideas out of the laboratory and start small companies in or outside the science park to move their ideas into the market place. Tapping-in involves firms without experience with the host university joining the park in order to take advantage of the expertise, facilities, and knowledge available at the university. Van Dierdonck *et al.* [53] and Massey *et al.* [31] suggest that these approaches may not be especially effective if the goal is formal collaboration with the host university. In the case of academic spin-offs, university regulations or meddling by administrators can hinder the creation of the firms [13, 49]. The Belgian study and a similar one conducted in the UK suggest that the majority of firms in a park that could be tapping-in are not and, conversely, they may be more likely to tap-out, that is, create collaborations with academics outside the park. Further, informal contacts may be far more numerous and significant between park firms and the host university than formal relationships [31, 53]. (Section 1.4. examines spin-off companies in greater depth)

Developed countries have learned that science or technology parks 'require two or three decades to reach their full potential and involve millions of dollars of investment' [26, p. 6]. Unlike other mechanisms for building industry–university links, science or technology parks take much longer to take shape and develop full operational capacity. This fact is not always appreciated or heeded [26]. As more of the university-based parks run up against political and economic problems due to the time factor, park officials and proponents are rethinking their estimates of how much time is needed for the parks to have impact. In the US, early park failures have made many state and university officials aware of the need to reduce expectations, especially when seeking support from state legislatures, which are sensitive to political pressure for quick results [24].

The time element, plus the large initial expense, often lead to financial exigencies becoming the driver of operational decisions and changes in direction before enough time has elapsed for the original design of a park to be fully implemented. Under these circumstances, changes in operations or direction can alter the nature of industry–university links, leading to difficulty in evaluating their impact, efficacy, and outcomes [26].

Finally effective linkages require active programmes in place to bring about interaction between university researchers and park tenants. Without such a mechanism, differences in orientation among tenants, researchers, and park managers will decrease the probability that interactions will occur on their own. Responsibility for bridging the gap should be in the hands of one individual on the park's management team [26].

In the US, the high point for development of regional economic development programmes focusing on R&D was the mid- to late 1980s. Because of the rapid growth in the number and types of such programmes, efforts were made to catalogue similar programmes across the country [33] and case studies were performed [44]. Starting in the late 1980s, a number of states began to experience economic downturns. Reductions in revenues and significant political changes in state-level legislatures reduced the scope of a large number of programmes, and even put whole state R&D offices out of business [27].

Prior to economic downturns state officials had justified the proliferation of science parks by citing such successes as Research Triangle Park, Silicon Valley in California and Route 128 in Massachusetts as evidence that states could stimulate their own economic development. Of these three examples, only RTP makes the intended point. State funds were critical for the venture to pay off; however, the California and Massachusetts examples occurred as a result of university–industry links that had no relation to any formal state-sponsored initiative. State-level recessions soon reduced the attractiveness of such expensive, long-term activities. More recently, Silicon Valley and Route 128 have proved to be vulnerable to local economic difficulties as well.

Linking large efforts like science parks to regional economic improvement carries its share of consequences. It has been difficult to determine the overall impact and effectiveness of regional technology-based programmes, especially in the terms politicians desire. While waiting long enough for programmes to have a chance to take effect is important, the situation is more complex. At the apogee of development of such programmes in the US, Peters and Wheeler [44] concluded that existing analytical methods were inadequate to provide answers to the main question of whether a particular programme or technology strategy affected economic development of a region. Since then, some progress has been made on the methodological front [c.f. 10, 16].

1.3. EXTENSION SERVICES OR CONSULTANCY CENTRES

The concept of an extension service, also known as a consultancy centre, is well-established in the US. Its principal function is to provide practical assistance and advice to farmers or businesses. The concept has its origins in the Morill Acts of 1862 and 1890. These pieces of legislation provided Federal land for individual states to use for the creation of higher education institutions with curricula in which agriculture and engineering were to be equal to science and classical studies. Institutions developed as a result of the legislation became known as land-grant colleges and universities.

Two additional pieces of legislation, in 1887 and 1914, provided that the land grant institutions establish experimentation stations and extension services to serve each state's farmers and businesses. In the case of agriculture, both flourished. For the most part, however, only the training component was developed for engineering. Neither the experimentation station nor the extension service aspect was implemented on a national basis [19]. The following example is significant not only because of its success, but also because its existence is an exception to the rule.

1.3.1. *US: Georgia Institute of Technology Extension Service*
In 1960, the Georgia Institute of Technology (Georgia Tech) established an Industrial Extension Division. After more than 30 years of operation, it is among the oldest and most successful of the handful of examples of industrial extension services in the US. A network of twelve regional offices around the state of Georgia are linked to a central office in Atlanta, which in turn is the point of contact with the Georgia Tech Research Institute and the university's engineering faculty. Extension agents, who have faculty appointments with the Research Institute, serve a wide range of industrial clients with a plethora of technical needs ranging from production line application to sophisticated computer-assisted design. Most projects undertaken by the extension service are short-term and 'involve manufacturing processes, facility and materials planning, methods improvement, or cost control' [19, p. 14].

The service is designed for small, technologically unsophisticated firms that need assistance in achieving greater economic and competitive equity. The agents are assigned to geographic regions and operate as generalists. When a request is more demanding in scope than a single agent can handle, researchers from the Research Institute are available to provide the necessary expertise. Rarely do extension services offer research services [19].

The Industrial Extension Division is expensive, labour-intensive, and difficult to operate. Typically, two agents serve as many as 500 companies in a given region. The Division pays for up to five days of their time per project. Any additional time is paid by the company in a contract negotiated with the agents. State support is low; agents and research staff must spend much time looking for and conducting contract research. In 1988, the entire extension system conducted $85 million worth of contract research and received only $3 million in state and other funding [19].

1.3.2 *Ghana: Technology Consultancy Centre*
Ghana established the technology consultancy Centre (TCC) in 1971 to bring expertise from the University of Technology, at Kumasi, to the productive sector, especially to informal and small-scale enterprises. The goal is to promote Ghana's industrial development. TCC services are provided by faculty members from a number of faculties, including agriculture, engineering, science, and social sciences. Over the years, the Centre has been converted from a traditional extension service to one that emphasises 'the development, promotion, and transfer of appropriate technologies to small-scale industries' [12, p. 64]. The Georgia Tech Extension Service was the model for TCC, and Georgia Tech personnel assisted with the Centre's development [1].

Early on, small firms and individual craftsmen approached TCC for technical and general business assistance. Some were looking for information, others for manufacturing techniques. A number of enterprises sprang up as a result of the early inquiries. Initially, however, the Centre experienced difficulty transferring technology to entrepreneurs from the informal and small-scale sector of the economy until it developed the Intermediate Technology Transfer Unit (ITTU). Established to assist clients who lacked the sophistication and education to work with formal organisations such as banks, ITTU provides a whole host of services, such as workshops demonstrating new manufacturing techniques, renting its own facilities or machinery to clients, and subcontracting entire or partial manufacturing orders to new entrepreneurs who are clients [12].

TTC provides services to large-scale firms and government agencies as well. Projects are usually larger in scope or require a greater level of technical expertise than those which are operated out of ITTU. Projects in which faculty members have consulted with large firms and government agencies have ranged in topic from road system design to development of new techniques for manufacturing brick. Clients have included the State Mining Corporation, an oil company, and the Ministry of Industries [12].

Some believe that TTC has not lived up to the expectations that many University faculty members had for it. The specific concern has been that the emphasis on appropriate technology and small-scale enterprise development diverts faculty members from part of the original function of receiving client inquiries and referring them to the appropriate University department. Faculty members feel that their real expertise is being underutilised, and that this reduces the effectiveness of the Centre [12].

1.4. SPIN-OFF COMPANIES

Spin-off companies were mentioned above in Section 1.2.3 in conjunction with science or technology parks. In fact, these firms can be both the outcome of specific models of IUR as well as a mechanism for IUR themselves. In the case of some of the models presented above, spin-offs result from industry–university research collaboration. For example, faculty researchers working in a US Engineering Research Centre may develop some type of intellectual property that has potential commercial value. If the researchers spin off the technology, idea,and so on and create a company to commercialise a product based on the discovery, then the centre has served as a mechanism to enable the precursor research to be conducted. The discovery that is spun off is the output of centre research, and the spin-off company formed to commercialise the discovery is an outcome of the centre's research.

In other instances, however, the creation of the company is itself a mechanism for IUR. This is the case when universities, regions, states, and so on establish schemes to establish spin-off companies independently of other IUR models.

The University of Twente in the Netherlands has established two programmes to create spin-off companies: the TOP Programme (Temporary Entrepreneur Places) and the TOS Programme (Temporary Support Spin-Offs). The former provides support for the University's graduates and researchers to start their own companies. TOS targets previously unidentified spin-off opportunities existing in multi-national corporations. From their experiences with TOS, the University went on to develop an aid programme designed to replicate TOS in developing countries [2].

The national origin of the spin-off company, or academic start-up, model is unclear. Uncertainty also exists as to whether this model is necessarily a good one to replicate. This is true with the models discussed previously. In a number of countries, there is little consensus regarding how closely linked research HEIs should be to academic start-up companies which have their origins in specific results from academic research. Underlying such uncertainty is the lack of consensus in many countries regarding how close the HEIs–industry relationships should be allowed to develop.

Spin-off companies provide a whole host of dilemmas for university administrators, faculty, and students in the US. The underlying concern is the fear that the priorities of the private sector will exert a major influence on the policies and activities of the university. Policy questions concerning facuity spin-offs abound, such as: (1) Is it acceptable for a student's thesis topic to be relevant to his or her faculty advisor's spin-off company? May students be supported by funds from their advisor's spin-off company? (3) Is it permissible for the university to have investment ties to a profit-making venture capital firm that is intended to support creation of faculty spin-offs? and (4) Is it permissible for faculty researchers to use university laboratory facilities to perform research for their spin-off companies?

In the US, there are plenty of examples of major universities having difficulty establishing workable technology management policies and apparatus. Frequently, international visitors to US universities with highly successful technology licensing and management operations, such as the Massachusetts Institute of Technology, University of Wisconsin at Madison, and Stanford University, conclude that the rest of the more than 3,000 US colleges and universities are equally successful and that there is a 'US model' of technology management and transfer. There are clearly excellent operations in US universities at both public and private institutions. Nonetheless, it is by no means the case that all universities are equally skilled and successful with technology management and transfer.

Some institutions are pushing for closer and closer ties to industry, at least partially to obtain more private sector research support. Others, however, are defining limits to such links. A few bad experiences in dealing with industry is all that some university policy makers need to cause them to restrict university–industry interactions. Problems can arise from such things as lack of experience in working with companies; underestimation of the time, care, and resources necessary for a successful spin-off venture; and lack of adequate diligence in negotiating intellectual property agreements [6].

2. Obstacles to Interaction

No two HEIs, let alone two countries, have precisely the same operational conditions and traditions, opportunities, and constraints. As a result, replicas of models of IUR that foster technology transfer are rarely exact copies of the original entities. To understand the factors external to the programme or activity that affect the need for adaptation, it is useful to consider general conditions that can hinder the development of effective IUR, and therefore technology transfer.

2.1. ACADEMIC TRADITIONS AND VALUES

Countries seeking to build the technological sophistication of their industrial sector frequently look to models in other countries that link colleges and universities with companies. However, faculty in countries where most research is conducted at research institutes rather than in HEIs often have little or nothing to offer the private sector unless the faculty members also have an appointment in a research institute. The separation of r esearch from higher education also discourages including students in joint research with industry when joint projects do occur.

When faculty conduct research in colleges and universities, the traditional missions, as well as value system and reward systems of western Europe that grew out of the Nineteenth Century German model of higher education, exert strong disincentives to interaction with industry. Academicians who receive their graduate training in countries whose colleges and universities operate according to this model may be particularly reluctant to engage in research geared to national or local needs. If through their training these researchers have come to value basic research and the pursuit of new knowledge for its own sake, the result can be particularly strong biases against involvement with real-world problems and interaction with industry [18].

In countries in which professors are government employees on 12-month contracts, regulations often prohibit them from earning extra income as consultants while conducting research for industry. Thus, even when industry is interested in making use of professors' research expertise, it may be illegal for research services to be remunerated. With faculty salaries being low from the start, this situation provides little incentive for faculty to interact with industry at all.

The 12-month government contract arrangement with no provisions for earning income through IUR reflects a lack of national tradition of IUR or awareness of its value. Fear of the effect of collaboration on traditional academic values often reinforces this view. Some faculty members and university administrators are afraid that IUR will endanger their institutions' basic research and graduate training missions [7, 8, 14].

Similarly, university researchers are sometimes concerned that engaging in industry-sponsored or applied research will, at minimum, be of no benefit, and might even hurt their careers [37]. The career constraint problem stems from the traditional reward structure placing little or no value on research with industrial or practical relevance. Another perceived threat to academic values is that industrial practices will restrict academic freedom, especially when intellectual property rights agreements include delays in publishing research findings that conflict with academic traditional patterns of the dissemination of knowledge [3, 37, 53].

Research results published in the last five years indicate that there has been a definite trend towards greater acceptance of IUR in US colleges and universities. Factors that affect likelihood of acceptance include field of study, prior institutional experience with IUR, and a desire to increase institutional prestige [9, 15, 45]. Recent findings by Lee [29] suggest that researchers in the mid-1990s are much more positively disposed to IUR than they were in the 1980s. Support for IUR is highest in disciplines that tend towards application and institutions in the lower quartiles of institutional prestige. While most patents issued and licenses granted are associated with the most prestigious institutions, faculty from these institutions are less supportive of IUR than those in institutions with lower prestige. As might be predicted, the predominant reason for opposition to close industry–university links is fear of degradation of traditional academic research values.

2.2. INDUSTRIAL PRIORITIES AND CULTURE

Companies of all sizes in all technology areas operate in an environment very different from that of universities. Similarly, industrial researchers function with expectations and priorities worlds away from those of their academic colleagues. In the US and many other countries as well, the 1980s saw changes in the industrial sector such that corporate managers had to face 'increasingly stringent demands for performance: continual quarter-by-quarter improvements in earnings, return on investment that [would] generate cash for growth, and price/earnings ratios that [would] keep asset valuations at high levels' [50, p. 26]. Time scales for R&D projects became tied to time scales in corporate strategic plans [21].

More recently, technology-based companies have felt additional pressures to be competitive in the global marketplace. At the same time, they have become more risk-averse and cash-poor. They are less likely than in the past to take chances in their selection of projects to pursue, and more likely to spend cash only when certain that the expenditure will result in added value than before. To make collaborations with HEIs worthwhile, companies expect researchers to be knowledgeable about industry's needs if they are to provide the necessary assistance [30].

There are differences between small and larger companies. Smaller firms continue to be the main source of innovation at the earliest stages of emerging technology development. Larger companies are strongest when working with more mature technologies [46]. Small technology companies have special problems securing funding for further development of emerging technologies. Venture capital is considered the natural source of funding for prototype development. However, in practice, venture capital is one of several sources of start-up capital. Additionally, venture capitalists are steering increasingly clear of small deals and technology firms. A number of analysts in the US and the UK have suggested the possibility that there is a dearth of funding options for early technology and company development [36, 51, 52].

The roles of entrepreneurship and willingness to accept risk are crucial for successful formation of new companies. The creator of the company must have a strong desire and the necessary skills to move the concept or product upon which the firm is based and must move the product from the idea stage to the marketplace before the firm can start collecting revenue from sales of the product. It takes money, time, and a lot of work to get to the point where there is a product to sell. Nonetheless, entrepreneurial spirit on the part of the company's founder is not enough. Those in the business of providing financing during the development of a product have other priorities. Specifically, they look for firms that are most likely to become very profitable in a short period of time. Financiers are most interested in keeping the risk of not receiving their funds back with acceptable interest in a short period of time as low as possible. As a result, they are not likely to back a company whose initial product is not likely to succeed, but would be very profitable if it did. Firms with moderate sales potential or good long-term prospects often find it difficult to obtain initial backing as well.

3. Beyond these Models

This chapter has touched on a number of different models for transferring technology, knowledge, and know-how between HEIs and private industry. There are many more [c.f. 2, 45]. No attempt has been made to judge the appropriateness of the adoptions described above, or, for that matter, the original models themselves. In order for a specific approach to have been transferred to another country, someone or some organisation must have liked what they saw, read, or heard about the original in order to try it in their country. Such choices can sometimes be made despite the lack of no consensus in the originating country as to whether the model has been 'successful', regardless of how success is determined. The intent here has been to present the cases without making judgements.

In the same spirit, it is useful to note in passing two other recent instances of adoption and adaptation of one country's approach to industry–university links by another country. One model that originated in Germany is of particular interest to some people in the US. The Fraunhofer Institutes are 'intermediate institutions' where early development is conducted on emerging technologies originating in universities and national research facilities that have commercial potential. Funded by the national Länder governments and private industry, the Institutes focus on projects that are considered beyond what universities and government research institutes should be doing but are viewed by individual companies as being at a stage that is still too risky for them to finance and conduct entirely on their own. Over the past several years, there has been consideration of establishing Fraunhofer-like facilities at several US sites to expand the capabilities of specific universities and respond to the needs of nearby companies. The US now has at least one such institute. The concept of a Fraunhofer Institute in the US has diverged somewhat from the original through closer links with higher education institutions than is typical in Germany.

A model that was tested in the UK in the early 1990s is already being tested in the US. Known as Technology Audit, this approach involves teams of technical experts who are familiar with the needs of particular companies performing an inventory of a variety of unrecognised opportunities for companies in university laboratories or research institutes, for example, technology, process and other discoveries 'sitting on the shelf' in laboratories; equipment with down-time; and specific personnel expertise. These opportunities are then matched with individual company needs or interests. In many respects, this approach resembles the initiative in the Netherlands that targets previously unidentified spin-off opportunities existing in multinational corporations described above [2].

Pilot tests of different methodologies for conducting the 'audits' were funded in 1992 by the UK Department of Trade and Industry (DTI). At a 1993 DTI-sponsored workshop designed as a debriefing of the pilot tests, a DTI official described the utility of Technology Audit this way: 'Firms need to benefit from the skills and knowledge available in our publicly funded HEIs. Technology Audit is the key to their proper identification and exploitation.... It should be an integral party of the research activities of HEIs' [20, p. 7].

Not long after the workshop, the US Navy and other US organisations entered into an agreement to conduct a pilot test of Technology Audit at a redundant military facility in Virginia. That test is now under way. While the test site is quite different from the university and research institute laboratories that were involved in the UK tests, the project team includes individuals who developed one of the methodologies that was tested under DTI auspices in 1992.

4. Observations

Looking across the breadth of models and patterns of replication, it is apparent that specific models can become 'fashionable'. One could argue that industry–university research centres are in this category now. Fashion and inter-country competition should not enter into decision making about adopting models of technology transfer. Regardless of how much politicians might advocate establishing centres, or whatever, countries are only inviting disappointment when they set out on a course that can never take them where they want to go. Few countries have the funds to waste on projects that have little or no possibility of being successful.

Everyone involved in making decisions about adopting or adapting technology transfer models must understand the time frame in which a particular model operates. Expecting a specific number of new jobs to be generated by the third year of a science park or a centre sets everyone up for a major disappointment. Those making funding decisions must understand and be willing to accept the amount of lead time that is required for a particular model to reach the stage of producing major results. Additionally, they must also understand what types of outputs and outcomes are reasonable and which clearly are not. While it is unlikely that such decisions can be made so cleanly, that political and financial support will be long-term, and that the nature of the likely outputs and outcomes is understood by all with a stake in the matter, the more decision makers consider these factors before any model is selected and a plan implemented, the greater the likelihood of eventual success.

In Section 1.2. above on science and technology parks, the role of governments (usually state and local) was mentioned. Regardless of the specific technology transfer model, governments are often significant players. In the US, their most important contribution is to provide the stimulus for participants to identify each other and come together. The initial motivation for such involvement is often government's willingness to provide at least partial funding for a collaborative venture. Beyond financial support, government contributes to success by influencing the conditions under which the links take place. Thus the governmental role is primarily that of a catalyst. University and industry partners actually make the collaboration come about [4, 37, 41].

Porter argues that governments can be helpful in promoting the conditions under which clusters of universities and businesses form partnerships. Options include tax incentives for businesses that encourage private investment in growth firms, policies that encourage lively domestic competition, and expenditures for roads, airports, and training programmes geared towards the needs of local industrial clusters [35]. In addition, tax credits for investing in industrial R&D and deductibility of new equipment donated to universities are viewed by many as significant stimuli for IUR [37].

Lederman [28] compares the government role in R&D across several other countries. He notes that the major economic powers are scattered across a continuum. At one end, there is the US approach, which he characterises as a pluralistic, less centralised, and market-oriented role. At the other end is France, with its centralised, planned, and strategically targeted role. The UK, the Federal Republic of Germany, and Japan are in between these extremes. Unlike the US, these other countries 'achieved a consensus some time ago that the central government has a clear responsibility to support S&T [Science and Technology] to serve civilian industrial needs. . . . In the US, no such consensus exists, and debate continues about such a strategy' [28, p. 280].

A related issue is the government role in financing private sector technology development. At least some individuals in the UK and US perceive a gap in the availability of capital for early technology development that is a function of the amount of risk technology-based firms and venture capitalists are willing to assume [36, 51, 52]. The ramifications of such a situation are ominous. Further research in this area would shed light on the extent to which there is a problem.

5. Final Thoughts

Models of IUR seeking to bring about technology transfer have been transposed and modified to suit specific situations for a long time. Adapting and translating—as opposed to copying—the original is necessary if models are to succeed in different settings. However, before one country looks to other countries for models of technology transfer to copy, the adopting country must make a thorough, honest assessment of specific needs that technology transfer can help address, and identify what role, if any, government should have in a technology transfer programme. In addition, it is also important to understand both the extent to which technology transfer is already taking place and existing obstacles to further transfer. Academic and industrial traditions, values, and priorities constitute only part of the range of possible obstacles to successful IUR and technology transfer. Taken together, a country's traditions and the relationships among different economic sectors affect the likelihood of any particular technology transfer model being successful in a new setting. As the examination of possible models takes place, underlying conditions in the originating country that contributed to success, plus conditions in the adopting country that mediate against successful adoption must be weighed. Finally, before the selection is made, it is wise to assess the political environment in terms of likelihood of sustained support—both political and financial—and acceptance of the long-term nature of such a venture.

Lessons from the past provide valuable guidance. When heeded, they can help ensure that the future is not a costly repeat of past disappointments. No country can afford such unfortunate and unnecessary mistakes.

Notes

1. Linda E. Parker is an Engineering Program Evaluation Director with the National Science Foundation.
2. Any opinions, findings and conclusions or recommendations expressed in this publication are those of the author and do not necessarily reflect the views of the National Science Foundation or the US Government.

References

[1] Behrman, J.N. and Fischer, W.A. (1980), *Science and Technology for Development: Corporate and Government Policies and Practices*, Oelgeschlager, Gunn & Hain, Cambridge, MA

[2] Bell, E.R J. (1993), *Some Current Issues in Technology Transfer and Academic Industry Relations—A Review Paper.* Paper Prepared for Oxford Trust, Oxford, England.

[3] Berman, E.M. (1990), 'The economic impact of industry-funded university R&D', *Research Policy*, 19, 349–55.

[4] Blais, R.A. (1990), *From Research to Production: Reflections on Technological Development Strategies and Relationships Between University and Industry.* Paper presented at the International Seminar on the New Political Context of Scientific and Technological Development, Montevideo, December 6–8, 1990.

[5] Blumenstyk, G. (1990), 'Pitfalls of research parks lead universities and states to reassess their expectations', *The Chronicle of Higher Education*, July 5, 1990, A19, A24.

[6] Blumenstyk, G. (1995), 'Turning off spinoffs', *The Chronicle of Higher Education*, July 21, 1995, A33, A35.

[7] Bok, D. (1990), *Universities and the Future of America*, Duke University Press, Durham, NC.

[8] Bollag, B. (1990), University-Industry Collaboration in Europe Called Mainly Positive and Likely to Expand, *Chronicle of Higher Education*, 7 November.

[9] Campbell, T. and Slaughter, S. (1995), *Protecting the Public's Trust: A Search for Balance Among Benefits and Conflicts in University-Industry Relations.* Paper given at the 1995 American Association for the Advancement of Science Annual Meeting, Atlanta, Georgia, 16–21 February, 1995.

[10] Centre for Economic Competitiveness (1992), *New York State Centres for Advanced Technology Program: Evaluating Past Performance and Preparing for the Future*, SRI International, Menlo Park, CA.

[11] Dahlman, C.J. and Sananikone, O. (1990), *Technology Strategy in the Economy of Taiwan: Exploiting Foreign Linkages and Investing in Local Capability*, World Bank, Washington, DC.

[12] Djangmah, J.S. (1992), *University Productive Sector Linkages in Ghana: Universities and the Small and Medium-Scale Enterprises.* Report prepared for the Association of African Universities.

[13] Etzkowitz, H. and Peters, L.S. (1991), 'Profiting from knowledge: organisational innovations and the evolution of academic norms', *Minerva*, 29, 133–66.

[14] Fairweather, J.S. (1990), 'Education: the forgotten element in industry–university relationships', *The Review of Higher Education*, 14, 33-45.

[15] Feller, I. (1990), 'Universities as engines of R&D-based economic growth: they think they can', *Research Policy*, 19, 335-348.

[16] Feller I. and Anderson, G. (1994), 'A benefit-cost approach to the evaluation of state technology-development programs', *Economic Development Quarterly*, 8, 127–40.

[17] Finnish Academy of Technology (1989), *High Technology from Finland*, Finnish Academy of Technology, Helsinki.

[18] Jones, G. (1971) , *The Role of Science and Technology in Developing Countries*, Oxford University Press, London.

[19] Jones, R.C., Oberst, B.S., and Lewis, C.S. (1990), 'Building US economic competitiveness: the land-grant model', *Change*, May/June, 11–17.

[20] Kingham, D., Ray, D. and Kirkland, J. (1993), *Commercial Opportunities for University Research—The Role of Technology Audit*, Oxford Innovation, Ltd, Oxford.

[21] Klimstra, P.D. and Potts, J. (1989), 'Managing R&D projects', in Industrial Research Institute, *Managing Research and Development—What We've Learned in the Past 50 Years*, Industrial Research Institute, New York, pp. 42–58.

[22] Korean Science and Engineering Foundation (1990), *KOSFF*, Korean Science and Engineering Foundation, Daejun City, Korea.

[23] Korean Science and Engineering Foundation (1991), *SRC/ERC 1991*, Korean Science and Engineering Foundation, Daejun City, Korea.

[24] Korean Science and Engineering Foundation (1994), *Science and Engineering Research Centre*, Korean Science and Engineering Foundation, Daejun City, Korea.

[25] Kyung, J-C. (1992), Director General for Manpower Policy, Ministry of Science and Technology, Korea. Personal Communication.

[26] Lalkaka, R. and Schiff, N. (1990), *Establishment of Technoparks in Turkey: Report of SPO/UNFSTD Preparatory Mission*, UNFSTD, New York.

[27] Lambright, W.H., Teich, A.H.. and O'Gorman, M.J. (1992), 'The turbulent condition of state S&T programs in the 1990s' in Lambright, W.H. and Rahm ,D. (eds), *Technology and US Competitiveness—An Institutional Focus*, Greenwood Press, New York, pp. 71–82.

[28] Lederman, L.L. (1994), ' A comparative analysis of civilian technology strategies among some countries: France, the Federal Republic of Germany, Japan, the United Kingdom, and the United States', *Policy Studies Journal*, 22, 279–95.

[29] Lee, Y.S. (1996), ' "Technology" transfer and the research university: a search for the boundaries of university–industry collaboration', *Research Policy*.

[30] MacLachlan, A. (1994), 'Industrial expectations and the research universities', *Research-Technology Management*, 37, 9–10.

[31] Massey, D, Quintas, P.R., and Wield, D. (1992), *High Tech Fantasies*, Routledge, London.

[32] McHenry, K.W. (1990), 'Five myths of industry/university cooperative research—and the realities', *Research-Technology Management*, 33, 40–42.

[33] Minnesota Department of Trade and Economic Development (1988), *State Technology Programs in the United States, Office of Science and Technology*, St. Paul, MN.

[34] Monck, C.S.P, Quintas, P.R., Porter, P.B., Storey, D. J., and Wyranczyk, P. *Science Parks and the Growth of High Technology Firms*, Croom Helm, London.

[35] Morgan, D. (1992), ' Think locally, win globally: Harvard's porter pushes regions clusters as the key to industrial competitiveness', *The New York Times*, 5 April, H1, H5–6.

[36] Murray, G.C. and Lott, J. (1995), 'Have UK venture capitalists a bias against investment in new technology-based firms?', *Research Policy*, 24, 283–99.

[37] National Science Board (1982), *University–Industry Research Relationships*, National Science Board, Washington, DC, NSB-82-2.

[38] National Science Foundation (1994), *Engineering Research Centres—Program Announcement,* NSF, Arlington, Virginia. NSF 94–150.

[39] National Science Foundation (1995), *Highlights of Engineering Research Centres Education Program*, National Science Foundation, Arlington, Virginia. NSF 95–.56.

[40] Office of Technology Assessment (1984), *Technology, Innovation, and Regional Economic Development,* US Congress, Office of Technology Assessment, Washington, DC. OTA-STI-238.

[41] Osborne, D. (1990), ' Refining state technology programs', *Issues in Science and Technology,* 6, 55–61.

[42] Parker, L.E. (1992), *Industry-University Collaboration in Developed and Developing Countries*, World Bank, Washington, DC, PHREE/92/64.

[43] Parker, L.E. (1995), *NSF's Research Centres: Government–University–Industry Partnerships*, Paper given at the 1995 American Association for the Advancement of Science Annual Meeting, Atlanta, Georgia, 16–21 February.

[44] Peters, L.S. and Wheeler, P.A. (1988), *Technology-Based Regional Economic Development: An Overview*, Centre for Science and Technology Policy, Troy, NY.

[45] Rahm, D. (1994), 'Academic perceptions of university–firm technology transfer', *Policy Studies Journal,* 22, 267–78.

[46] Roberts, E.B. (1989), 'Managing invention and innovation', in Industrial Research Institute, *Managing Research and Development —What We've Learned in the Past 50 Years*, Industrial Research Institute, New York, pp. 3–21.

[47] Science Park Administration (1989), *Questions and Answers*, Science Park Administration, Taiwan.

[48] Segal, N.S. (1987), 'IHE links: the need for institution-specific approaches', *Industry and Higher Education.*

[49] Segal Quince Wicksteed (1988), *Universities, Enterprise and Local Economic Development: An Exploration of Links*, HMSO, London.

[50] Steele, L.W. (1989), 'Selecting R&D programs and objectives', in Industrial Research Institute, *Managing Research and Development—What We've Learned in the Past 50 Years*, Industrial Research Institute, New York, pp. 22–41.

[51] Thomson, H. (1993), 'Commercial opportunities for university research', in Kingham, D., Ray, D. and Kirkland. J. (eds.), *Commercial Opportunities for University Research—The Role of Technology Audit,* Proceedings from a conference and workshop dealing with Technology Audit. December 2–3, London, pp. 63–71.

[52] United States Department of Commerce and Department of Treasury (1992), *Financing Technology—A Report of the Financing Technology Roundtables,* United States Department of Commerce and United States Department of Treasury, Washington, DC.

[53] Van Dierdonck, R., Debackere, K., and Engelen, B. (1990), 'University–industry relationships: how does the Belgian academic community feel about it?', *Research Policy*, 19, 551–66.

GPSR Compliance

The European Union's (EU) General Product Safety Regulation (GPSR) is a set of rules that requires consumer products to be safe and our obligations to ensure this.

If you have any concerns about our products, you can contact us on ProductSafety@springernature.com

In case Publisher is established outside the EU, the EU authorized representative is:

Springer Nature Customer Service Center GmbH
Europaplatz 3
69115 Heidelberg, Germany

Batch number: 10815974

Printed by Printforce, the Netherlands